PRAISE FOR ANTHONY BENNETT

How to Survive Like the Amish is a practical manual—skills to feed, purify, preserve, and protect your family when the lights go out.

GRACE WHITAKER, INDEPENDENT LIVING ADVOCATE

The book emphasizes local networks and hands-on know-how. It builds resilience from the ground up.

JONAH REED, RURAL COMMUNITY BUILDER

Clear, doable steps, durable tools, and a community-first blueprint that works in real life.

DR. LENA HART, SUSTAINABLE LIVING ADVOCATE

HOW TO SURVIVE LIKE THE AMISH

HOW TO SURVIVE LIKE THE AMISH

A PRACTICAL, LOW-TECH BLUEPRINT FOR STAYING FED, WARM, AND CONNECTED WHEN THE GRID GOES DOWN

SELF-SUFFICIENT LIVING

BOOK 4

ANTHONY BENNETT

Published by Book Bound Studios

First Edition

To those who plan, practice, and stand ready when the grid fails.

I went to the woods because I wished to live deliberately, to front the essential facts of life, and see if I could not learn what it had to teach, and not, when I came to die, discover that I had not lived.

HENRY DAVID THOREAU, WALDEN, 1854

CONTENTS

INTRODUCTION

THE HOUSE GOES QUIET FAST

Modern comfort rests on a thin stack of moving parts. Electricity, trucking, payment systems, water treatment, fuel delivery, refrigeration, cell towers, replacement parts, and the quiet assumption that someone else will keep all of it running. When even one layer falters, a normal house changes character in a hurry.

A long outage does more than turn off the lights. Frozen food starts softening, well pumps stop, furnaces go cold, freezers sweat, toilets become a problem, and every task that once took seconds begins demanding time, fuel, labor, and judgment. People who felt prepared because they had a flashlight and a case of bottled water learn how much of life depends on systems they never notice.

Recent years have made that plain. Ice storms have shut down power for days. Wildfires have forced utility cutoffs. Floods have contaminated wells. Cyberattacks have interrupted fuel and food distribution. Inflation has turned routine pantry items into budget decisions, and supply chain gaps have left shelves empty of simple things like canning lids, feed, and medicine.

Most households were built for convenience, not interruption. They run lean on stored food, low on repair parts, short on practical skills, and heavy on habits that assume stores will stay open and taps will keep flowing. A family can own expensive gadgets and still be fragile if its daily life depends on electricity, delivery trucks, and one-click replacement.

That fragility creates a certain kind of fear. It is the anxious feeling that comes from knowing your household works beautifully under normal conditions and poorly under stress. People sense the weakness before they can name it, which is why many drift toward buying gear: batteries, radios, solar chargers, freeze-dried meals, water filters, bins, kits, and backup devices that promise security in a box.

Some of that equipment helps. Much of it becomes clutter because the problem was never lack of stuff alone. The real problem sits deeper, in the difference between a house that *consumes* and a house that can still *function* when outside support goes missing.

That difference is the reason this book exists. It begins with a blunt fact: households that already know how to cook from scratch, store staples, mend clothing, preserve food, haul water, heat one room, use hand tools, and rely on nearby people stand on firmer ground during disruption. Those habits matter in a weeklong blackout, in a bad winter, in a stretch of layoffs, and in a drawn-out failure where no one can say exactly when things will return.

You do not need to accept every prediction about collapse to see the value here. A family that keeps a working pantry, a way to cook without the grid, a backup water plan, extra blankets, repair skills, and strong ties with neighbors is harder to shake. That household spends less, wastes less, panics less, and recovers faster.

Preparedness has suffered from two bad sales pitches. One turns it into entertainment filled with dramatic scenarios, expensive tools, and chest-thumping. The other turns it into vague lifestyle advice with mason jars, charming photos, and no serious plan for cold weather, sickness, sanitation, or hard labor.

A useful approach sits in the middle. It deals with ordinary needs under abnormal conditions: breakfast when the stove is dead, clean clothes when the washer is useless, safe water when the tap runs brown, warmth when central heat quits, and repair when no service truck is coming. It asks how a household keeps working, day after day, when convenience disappears.

That question points toward people who have practiced versions of this for generations. Not because they are frozen in time, and not because they are immune to hardship, but because their daily life has kept many old strengths alive. They have experience with low-energy living, durable goods, local trade, hand labor, and community support that modern households badly need to study.

The Amish have spent a long time building homes that can still function when the wider machine stumbles. That does not make them mythical, and it does not make them perfect. It makes them useful teachers for anyone who wants a house that can keep feeding, heating, cleaning, repairing, and carrying on when the grid fails.

BORROW THE PLAYBOOK, NOT THE IDENTITY

The Amish attract attention for obvious reasons. They use horse-drawn transport in many places, dress plainly, limit some technologies, and organize life around church districts, family labor, and strong local ties. People often stop there, treating them as a curiosity or a symbol.

That misses the point that matters for survival. Their strength comes from *systems*: food grown near home, workshops built around hand skills, households trained to work with fewer appliances, and communities that expect members to help one another through illness, fire, crop loss, and everyday need. Those patterns create resilience long before any crisis arrives.

This book is not an argument for becoming Amish. It is a guide to learning from practices that hold up under stress. You do not need a buggy, plain dress, or a shared theology to understand the value of root cellars, pressure canning, wood heat, hand tools, repair culture, bulk staples, and a neighborhood where favors and labor still move in both directions.

That distinction matters because people often romanticize simple living. They picture candlelight, homemade bread, and peaceful evenings while skipping over hauling water in winter, splitting wood in sleet, scrubbing laundry by hand, cleaning up after animals, and fixing a broken gate before sunset because tomorrow depends on it. Real self-reliance carries weight, dirt, repetition, and discipline.

The Amish example is useful precisely because it is practical. Their households often build routines around necessities instead of trends. Many know how much flour, lard, vinegar, kerosene, feed, fabric, seed, and firewood it takes to carry a family through a season because they have measured those needs in work and storage, not just in shopping trips.

They also tend to favor goods that can be repaired. A cast-iron pan, a treadle sewing machine, a hand pump, a work-horse, a wagon, a woodstove, a sturdy table, a line of jars in the cellar, and a cupboard of soap, salt, oats, and beans can look old-fashioned to people trained to admire sleek convenience. In a disruption, those things look like common sense.

The lesson is not to reject every modern tool. The lesson is to judge tools by their dependence, durability, fuel needs, and failure points. A household grows stronger when it keeps what works well, adds low-tech backups, and builds routines that do not collapse when the power, internet, or supply chain drops away.

There is another reason the Amish model deserves attention: it ties self-reliance to *interdependence*. Many modern people imagine preparedness as a lone household with locked doors and private stockpiles. Amish communities have long shown a tougher model, where families maintain their own competence while staying woven into a network of labor, exchange, trust, and obligation.

That social structure matters as much as the pantry. One family may own a sawmill, another keep bees, another have skill with births, another know harness repair, roofing, butchering, or canning large volumes of produce. No single house needs to master everything at once when the community has depth.

Most readers will adapt this in a mixed way. You may live in a suburb with a quarter-acre lot, a garage, and strict zoning. You may be on five rural acres with room for chickens and wood storage. You may have children, aging parents, a full-time job, a church network, a pickup, a bicycle, a shed full of tools, or none of the above.

That is fine. This book treats Amish practice as a working blueprint, not a costume. The goal is to extract the parts that strengthen any household: fewer dependencies, better routines, stronger hands, better storage, more repair, more local production, and steadier relationships with people close enough to matter.

Preparedness gets real when it leaves the fantasy aisle and enters the kitchen, the yard, the wash line, the woodpile, the workshop, and the neighbor's driveway. That is where useful resilience has always lived.

BUILD A HOUSEHOLD THAT CAN KEEP WORKING

A prepared household does more than store supplies. It builds repeatable ways to secure food, water, heat, sanitation, clothing, medicine, light, and repair with limited outside help. That shift from stockpiling to *systems* changes everything.

Think of the average kitchen during a blackout. A freezer full of meat looks reassuring until you realize you cannot cook it all, keep it cold, or preserve it fast enough. A pantry stacked with random cans offers calories, yet meals still become awkward if you lack a hand can opener, clean water, fuel, ways to bake, or recipes built around basic staples.

Competence starts where routine meets need. A family that regularly cooks beans, rice, oats, potatoes, soups, stews, bread, and simple skillet meals from stored ingredients can pivot faster than one whose pantry depends on frozen entrées and condiment-heavy convenience food. The same pattern holds across the whole house.

Water is not only about thirst. You need it for cooking, washing hands, cleaning dishes, managing waste, tending animals, laundering clothing, and treating minor illness. A plan that covers drinking alone will fail under the first week of real use because household life keeps demanding more.

Heat works the same way. Owning a portable heater means little if you lack safe fuel storage, ventilation, carbon monoxide awareness, sleeping arrangements for one warm room, layered bedding, draft control, and habits that conserve body heat and labor. Winter punishes sloppy planning fast.

Sanitation is where many people lose their nerve. When toilets stop flushing and washing machines sit dead, a home can turn unpleasant and unsafe within days. Cleanliness becomes work measured in buckets, soap, boiling water, drying racks, handwashing discipline, and careful handling of waste, gray water, diapers, menstrual supplies, and sickroom laundry.

Repair skills deserve the same respect as food storage. A snapped handle, torn coat seam, broken hinge, leaking boot, clogged stovepipe, weak fence, cracked bucket, or failed gasket can escalate from annoyance to serious trouble when stores are closed and replacement parts are distant or costly. Households that fix things stay calmer because small failures stop cascading.

The Amish way of life trains these responses through repetition. Children grow up seeing work divided, tools maintained, food preserved, clothing mended, gardens planned, and neighbors called in when a barn roof must go up fast. A modern family can borrow that pattern by making resilience part of normal life rather than reserving it for emergency talk.

That means practicing on ordinary days. Cook a meal without electricity. Wash a load of clothes by hand. Spend a weekend heating only one room. Keep a month's worth of staple meals and actually eat from them. Repair something before replacing it. Learn how much lamp oil, flour, soap, feed, wood, and drinking water your household truly uses.

These exercises do more than reveal gaps. They replace vague worry with measured knowledge. You learn which tools matter, which purchases were foolish, which skills need work, and which comforts were secretly dependencies.

This book is built for readers who want exactly that kind of grounded confidence. Some chapters will feel familiar if you already garden, can food, or split firewood. Others may challenge habits that seem harmless until you notice how many daily tasks depend on a stable grid, cheap fuel, abundant retail stock, and a service economy that can vanish under pressure.

You do not need to master everything at once. You need to build a home economy that carries more of its own weight, wastes less, stores wisely, and teaches every member how to contribute. **Capability grows through repetition, not excitement.**

STRONG HOUSEHOLDS NEED STRONG TIES

A family can store food, cut wood, repair tools, and still hit a wall. Illness spreads. Wells fail. Storm damage demands more hands than one household has. A calf gets out, a chimney needs work, a freezer full of meat must be processed today, an elderly neighbor needs hauling help, and a mother with three young children cannot keep every task moving alone.

The popular image of survival centers on isolation. Board the windows, guard the supplies, trust nobody. That picture ignores how humans have survived hard times across history: by pooling labor, sharing knowledge, trading useful goods, and building reputations for competence and fairness.

Amish communities understand this at a level many modern neighborhoods have forgotten. Mutual aid is built into daily life, not reserved for disaster headlines. Barn raisings are the famous example, but the deeper pattern is simpler and more demanding: people know one another, expect to help, and expect to be helped.

That expectation changes how resources are used. One household may not need every specialized tool if trusted neighbors can borrow and lend. Families can coordinate planting, butchering, canning, hauling, child care, transport, and elder support. Skills spread by watching, doing, and teaching rather than by scrolling through clips during a crisis.

Preparedness without community often turns brittle. It depends on one income, one body, one set of skills, one property line, and one pile of supplies. A resilient network has overlap. If your hand pump fails, someone nearby knows how to rebuild leather seals. If your stove pipe cracks, someone has sheet metal tools. If your child spikes a fever, someone has experience, herbs, supplies, or the vehicle fuel to help.

That does not require joining an intentional community or surrendering privacy. It requires becoming useful and trustworthy where you already live. Share produce. Loan tools and return them clean. Learn who keeps bees, who hunts, who cans, who welds, who bakes bread, who knows first aid, who has a chainsaw mill, who can sharpen blades, who can watch children for an hour while a roof patch gets done.

Families matter inside the house too. Children can learn chores that carry real value. Teenagers can cook, stack wood, patch fences, collect eggs, weed rows, and help with younger siblings. Older adults often carry the most precious survival asset of all: memory of methods that save money, stretch food, and keep morale steady when work is repetitive and conditions are rough.

One reason this book emphasizes routines is that routines reduce chaos. A household where everyone knows how water is stored, which foods are eaten first, how laundry gets washed off-grid, where candles and matches belong, how waste is handled, and who checks on which neighbor will fare better than a house where one anxious adult tries to manage everything alone.

The chapters ahead will ask you to examine your home with clear eyes. How do you cook if the stove is dead for a month? Where does your water come from if taps fail? How do you keep waste from becoming a health threat? What can you repair by hand? What can you grow, preserve, barter, or make last? Who would come if you truly needed help, and whose door would you knock on without shame?

Those are not abstract preparedness questions. They are household questions, and households answer them through habit. A calm, productive, low-dependence home offers a form of security that no shopping spree can buy.

When the grid fails, the winners will not be the people with the flashiest gear. They will be the households that know how to work, how to store, how to fix, how to share, and how to keep living in an orderly way when the larger system goes quiet.

ONE
THINK IN SYSTEMS, NOT SUPPLIES

FROM STOCKPILES TO ROUTINES

Panic buys fade fast. What keeps a family alive when the shelves are empty is not the size of a pantry but the habits that sit behind it. A real readiness comes from routines that run the day, not from a shopping list you rarely update. When a storm stalls delivery trucks or a blackout silences the fridge, your home should keep moving because every essential function has a built-in, practiced path.

A system is a chain of routines that holds together even if one link suffers. Start with the core needs: food, water, heat, sanitation, work, and cooperation. Each need has a practical method that stays usable without electricity or digital control. This is not about collecting gadgets; it's about repeating reliable tasks until they feel automatic. The Amish model isn't about abstaining from modern life. It's about letting simple methods carry you through disruption without panic.

Consider the day a power outage lasts beyond a few hours. A household built on routine uses stored staples not as a one-off stockpile but as a rotating pantry. A weekly schedule ensures you eat well without rushing to the supermarket at the last minute. A small wood stove or a safe outdoor heater can keep milk from spoiling and soup on the stove without a gas line. Laundry happens by hand in a tub, not by a push of a button. Rainfall, not a distant water main, becomes the water source for cleaning and cooking. In short, a steady rhythm replaces impulse buying with dependable practice.

This chapter helps you shift from chasing deals to building a living framework. It begins with the kitchen, the most visible signal of readiness, and expands to other daily loops: water you can fetch, heat you can generate, waste you can manage, and work you can keep doing without a grid.

First, stabilize the pantry with stored staples that travel well and store compactly. Choose a core list: dried beans or lentils, rice or cornmeal, oats, flour, canned tomatoes, dried fruit, canned fish or meat, salt, oil, and a few spices. Rotate routinely so yesterday's stock becomes today's meals. A simple rotation—FIFO, first in, first out—keeps spoilage from creeping in and saves cash when prices spike. Tie your rotation to a basic weekly menu that uses the same core ingredients in multiple ways. It's not exciting, but it's sturdy. A few flexible recipes can turn dull staples into satisfying meals: a pot of beans with greens, a pot of stew that uses left-overs, a pancake breakfast with pressed apples from last season's canning.

Storing water becomes part of the daily routine, not a separate panic project. A family that keeps several days of drinking water on hand makes every other precaution possible. Store it in food-grade containers, labeled with dates, and keep a plan for rotating it every six to twelve months. Clean water for cooking and sanitation should mirror that rotation. If you live in a dry climate, add a small rain barrel and a simple gravity-fed system to carry water to the kitchen sink or a wash station. In a crisis, your water plan reduces the risk of dehydration and makes sanitation feasible without municipal systems.

Heat and light form the second spine of a resilient home. A well-insulated space, layered clothing, and bedding make cold more bearable than it looks on paper. A wood stove or a safe wood-burning heater becomes a reliable backbone for warmth, hot water, and even light when the power is out. If wood isn't available or practical, a propane camp stove kept in good condition can fill gaps for limited durations. For lighting, candles and oil lamps are straightforward, low-tech options. A handful of solar-powered lanterns or a charged battery pack provides a buffer for critical tasks after dark. The goal is warmth and visibility without depending on an outage-proof electricity grid.

Sanitation remains non-negotiable. A clean home is healthier, and a simple waste system keeps odors in check and pests away. A basic bucket toilet with a liner and a sealant for odor control, paired with a handwashing station, keeps hygiene routines intact. A plan for laundry that preserves fabrics and keeps clothes usable—rub-a-dub tubs, a scrub board, and air-drying on a rack—reduces the reliance on electric washing machines. These routines are not glamorous, but they prevent a quick slide into illness and discomfort when the system stalls.

In practice, a practical household builds a rhythm of minimal disruption. It begins with one area—food—and expands to the rest of daily life, always with a view to finishing the day with a sense of control rather than panic. A stable home treats disruption as a temporary condition, not a catastrophe.

A system, once learned, becomes the default. It makes daily life predictable, and predictability is what keeps fear at bay when the lights go out.

MAPPING YOUR HOUSEHOLD DEPENDENCIES

A grid outage exposes the hidden architecture of your daily life. Some parts crumble quickly; others hold steady because they function without power, doors can be closed properly, or because they rely on manual processes. The first step is to name every critical function of the house and then identify everything that makes each function possible.

Begin with the big six: food supply, water supply, heat, sanitation, work/communication, and mobility. Then add the supporting roles that make those six possible: electricity for refrigeration, gas for cooking, fuel for heating, internet and phone for coordination, and the ability to transport essentials to and from the home. Each function should have at least one manual alternative. The goal is redundancy that does not require you to own a store of gadgets.

To map effectively, draw or list how your home delivers each function today. For food, list the fridge and freezer, pantry, and meals planned from core ingredients. If the power goes out, what remains usable? Do you have canned goods, dry staples, or preserved foods that can be prepared with hot water or fire? Do you rely on a farmers' market delivery schedule or a weekly grocery run that could be disrupted? For water, catalog where water comes from, how it's stored, and how it's moved to the kitchen. If the well pump stops, is there a gravity-fed reservoir, rain barrels, or a stored supply you can access for several days? For heat, note your primary heating source and any backup options. If you rely on electricity for space heating or water heating, what is the manual alternative? For sanitation, identify the toilet arrangements, waste disposal methods, and any interim solutions you can deploy quickly.

The exercise also reveals invisible lines of weakness. A modern home often depends on digital systems for monitoring, security, and communications. If the internet or cellular networks go down, you should still be able to manage basic coordination and safety. A simple printed map of the house, with care instructions for children and guests, becomes invaluable when screens fail. Keep a written checklist of daily tasks that can proceed without power; it's surprising how fast the pattern of work supports itself when you don't have to think through every step.

A dependency map isn't a static chart. It's a living document that you update after experiments, not just after a crisis. The act of mapping changes behavior: you'll start stocking different items, you'll add a new backup method, and you'll reallocate space for critical goods. The result is a clearer sense of what to protect and where to build resilience without purchasing more gadgets.

A practical map translates into clear roles. Assign a lead for water, a lead for food, a lead for heat, and a lead for sanitation. When everyone knows who handles what, daily routines keep moving even if someone is temporarily unavailable. The chart of dependencies becomes a household constitution—simple, actionable, and enduring.

BUILDING LAYERS OF BACKUP

A single backup fails when you need it most. The plan rests on multiple layers, each designed to cover a different time scale and a different failure mode. The first layer is everyday reliability: a small, practical backup you can implement without major changes, such as a well-organized pantry, a water container you refill weekly, and a commitment to cooking without electricity at least once a week.

The second layer is standby resilience. It includes practical alternatives that you can scale with your means: a wood stove or fireplace for heat and cooking, a hand-pump backup for water, a supply of candles and oil for light, and a durable, easy-to-clean clothing system that can be worn for days if washing is limited. The third layer addresses longer outages. This layer adds solutions that take time or effort to assemble but deliver value when the grid stays off for weeks: a small solar panel for essential charging, larger rainwater storage, a root cellar for seasonal produce, and a communal support plan to share labor and resources with neighbors.

Every area of life benefits from layering. Food becomes a series of buffers rather than a single point of failure. Stock your pantry, but also grow a few reliable staples in the garden. Learn simple canning or drying methods to extend harvests. In water, preserve extra containers, learn to collect and store rainwater, and practice purification methods such as boiling or solar pasteurization. Heat becomes a mix of wood, a simple solar heater if you have sun, and extra blankets for cold rooms. Lighting becomes a triad of candles, oil lamps, and a small, well-maintained solar lantern or battery-pack system. Sanitation uses a bucket system with a composting component, plus a basic plan for maintaining and cleaning clothing that preserves the integrity of fabrics.

Manpower matters. Shared labor within a household reduces strain and builds confidence. A neighborly exchange of skills—sewing, carpentry, repair, or food preservation—extends the reach of your backup layers. The more you lean on practical, non-electric methods, the more resilient your home becomes. Layering backups isn't about collecting gear; it's about weaving reliable options into daily life so that if one thread snaps, others hold the fabric together.

The ninety-day home audit

A practical ambush plan beats guesswork every time. The ninety-day home audit translates the dependency map and backups into a concrete, repeatable action plan. It starts with a simple assessment, then moves into implementation, practice, and refinement. The aim is not a perfect blueprint but a living, tested routine you can maintain.

Begin with a clear, three-part assessment: inventory, vulnerability, and priority. Inventory lays out what exists: foods, waters, fuels, tools, and the fundamental services you rely on. Vulnerability asks where you would fail first under stress: a failed well, a broken freezer, a flood-prone bathroom, a broken stove. Priority ranks fixes so you spend your time where it matters most.

From there, translate findings into a 90-day schedule. Week 1 and 2 are mapping and stocktaking. Week 3 and 4 focus on basic backups for food and water. Weeks 5 and 6 add heat and light backups. Weeks 7 and 8 test sanitation routines and laundry methods without power. Weeks 9 and 10 expand to minor repairs, tool maintenance, and basic repairs you can perform without electricity. Weeks 11 and 12 tighten routines, train family members, and solidify neighborly support arrangements.

A completed audit yields a practical action list you can work through without guesswork. It creates momentum and clarity. The household that finishes the ninety days with functional backups, practiced routines, and a shared sense of responsibility will stand a better chance when the grid goes quiet for weeks.

Each section of this chapter builds toward one simple truth: resilience is the product of habits, not purchases. A well-practiced system keeps a household steady when external systems fail.

TWO
FOOD SECURITY STARTS IN THE YARD AND PANTRY

The most reliable shield against weeks without power is a pantry that works without electricity.

Your kitchen is a machine you can make run on wind, water, and a bit of discipline. When the grid fails, the first thing to fail isn't the fridge; it's the mind. People panic, then meals vanish with a sigh. A well-tended pantry and a yard that keeps giving you food will steady the nerves and keep a family fed long after the lights go out. This chapter treats food security as a craft you can practice now, not a dream you hope to wake up to later. It's about concrete systems you can test, month after month, until they feel automatic.

SURVIVAL PANTRY

The pantry is where hunger is defeated before it ever starts. It's not a gadget collection; it's a deliberate library of calories, nutrients, and flavor that travels well in heat, cold, and neglect. A resilient kitchen starts with broad, shelf-stable foundations and builds up from there. Think grains, beans, fats, salt, and a few long-lasting staples that turn into meals with minimal fuss.

Grains form the backbone. Rice, oats, and cornmeal are versatile laborers that stretch a pot of soup, a pan of fried vegetables, or a pot of porridge into many meals. Flour isn't just bread; it becomes dumplings, dumplings become wrappers for leftovers, and a simple roux thickens a pot of beans into a meal all by itself.

Beans and lentils are the other half of the core. They're quick to rehydrate, nutrient-dense, and forgiving. Storing dried beans in airtight containers keeps them usable for years if kept out of direct heat and moisture. For every cup of dry bean, you get several cups of cooked protein—without refrigeration, without electricity, without fear.

Fats matter more than you think. Oil in a bottle, plus a sturdy animal fat like lard or tallow if you can source it, carries flavor and calories when many fresh ingredients fail. Fats can extend the life of a meal by making simple starches feel satisfying and finish a pot of beans with gloss and body.

Salt is a small thing with outsized impact on appetite and preservation. A salty flavor can turn a bland pot of beans into something comforting, and salt helps preserve other ingredients in a pinch. Sugar or honey acts as both energy and a shield for flavor when fresh fruit is scarce.

Canned goods and shelf-stable jars ride at the front of the shelf. Tomatoes in cans, tomato sauce, canned fish, and dried soups give a kitchen backbone when fresh greens are scarce. Powdered dairy—milk, eggs, or cheese—can boost nutrition when real dairy isn't available, though you'll use it sparingly due to cost and storage considerations.

Spices and dried herbs aren't vanity; they're value. A few jars of dried onion, garlic, bay leaves, cumin, paprika, and dried parsley turn dull meals into something nourishing and palatable. In a long outage, flavor matters as much as calories.

Rotation is the guardrail here. First in, first out. Use the oldest items first, replace with fresh when you have dollars and time to spare. A simple calendar on the pantry door can track what you opened last and what you need to restock before the next lean spell.

Stocking guidance should fit your family and your climate. A family of four aiming for stability during a several-month disruption should think in meals rather than items: enough grains to sit with a bowl of porridge, enough beans to pair with rice several times a week, enough canned tomatoes to season soups, enough oil to carry a pot and a pan. The exact pounds or cans aren't magic; the rhythm is. When the pantry is laid out for meals you already know you can cook, stress falls and decisions slow down.

A practical starter list borrows a few dozen items you can rotate. Think: rice, pasta, oats, flour, cornmeal, dried beans, lentils, canned tomatoes, tomato sauce, canned fish or meat, powdered milk or evaporated milk, cooking oil, salt, sugar, honey, bouillon or stock cubes, dried onions, garlic, a handful of herbs, and a couple of shelf-stable vegetables in jars or cans. Add a spice trio you actually use: a pinch of cumin, a dash of paprika, and a hint of dried thyme.

In a pinch, you can feed a family for weeks with those basics if you cook with care. But you should tailor the pantry to what your kitchen handles and what your family actually eats. If your household loves soup, keep more dehydrated vegetables and broth powder. If you bake, keep extra flour and sugar. If you rely on beans, stock more salt and oil to carry flavor and texture.

A concise inventory helps you see gaps and opportunities at a glance. A practical example for a family of four might include several categories: long-grain rice and rolled oats; a couple of pasta shapes; four to six varieties of dried beans; a bag of flour and a bag of cornmeal; sugar and salt; a small jar of oil; a few canned tomatoes, one or two cans of fish, and a couple of shelf-stable cheeses or powdered dairy; a set of spices; and a range of dried vegetables or tomato paste. You don't need every item every month; you need enough continuity to cook from memory without a trip to the store.

Once the pantry is organized, cooking becomes a predictable process. You build meals from what's on the shelf, not what's on special. You learn to pair staples into sustaining dishes: beans and rice, lentil soup with tomatoes, vegetable stews thickened with flour, porridge enriched with a spoon of oil and a pinch of salt. The goal isn't culinary ambition; it's reliability. A steady pantry makes sense of days when the grid fails and fear climbs.

The survival pantry isn't glamorous. It's steady, well-supported, and easily taught to kids. It creates a baseline of calories, texture, and flavor with minimal waste. When you can reach into a cabinet and assemble a nourishing meal without logging on or rushing to a store, that's resilience you can feel.

GROWING HIGH-VALUE CROPS

A kitchen that feeds a family through long disruptions hinges on crops that yield calories with modest effort. Grow what pays you back with minimal inputs, and you'll see the difference in the number of meals you can put on the table from your own soil. The emphasis is on calorie crops, dependable vegetables, and herbs that stretch flavor across weeks of cooking.

Potatoes top the list for a reason. They're dense, versatile, and forgiving. A small potato patch can keep a family well-fed through winter when other greens disappear. Sweet potatoes extend the supply of energy and store well in cool, dark spaces. Winter squash adds heft to soups and stews, and pumpkins or gourds store for months with very little fuss. These crops survive rough weekends, heat waves, and neglect better than most greens. They are the backbone of a stubborn kitchen that refuses to quit when stores run dry.

Beans and corn pair well with potatoes and squash, amplifying the pantry's reach. Dry beans store for years if kept dry and cool. Corn, ground into meal or popped as corn, offers energy and texture that keeps meals interesting. Onions and garlic sit near the base of any garden plan; their flavors carry through weeks of cooking and help stretch every other ingredient.

For leafy greens, pick hardy varieties that tolerate cool or warm seasons depending on your climate. Kale, collards, and cabbage manage cooler months and stay hearty when temperatures drop. Carrots, beets, and parsnips grow in dirt that stays cool beneath mulch, providing late-season nutrition when fresh food is scarce. Finally, cultivate a few kitchen herbs close to the kitchen door: thyme, oregano, dill, and parsley. They turn plain meals into something comforting without much effort.

Think in terms of harvest windows and storage. Potatoes, squash, and onions can keep in a root cellar or a cool, dark corner. Tomatoes and peppers stored in jars or dried become sauces for late winter, and beans dry down into a pantry staple that's easy to rehydrate for a quick meal. You'll want space to cure and dry as you harvest: a screened porch, a shed, or a wooden rack in a cool area of the basement.

Planning begins with a simple question: what can I keep here, and for how long? This question reframes gardening from a hobby into a hedge against hunger. A few well-chosen crops with steady yields will bear the burden when the market falters and shelves sit empty.

The most reliable approach is a mixed strategy: a few potatoes and storage roots, a compact bed of squash, a modest bean patch, and a couple of compact greens that harvest progressively. Even tiny gardens can deliver meaningful meals when you learn to stretch every square foot with careful planning and a good sense of season.

Seasonal planning matters as soon as soil thaws. Map out what to plant in early spring for a late-spring harvest, what to plant for summer meals, and what to harvest in fall for winter storage. You'll keep flavor intact while extending your ability to feed a family without frequent trips to the store.

The calendar becomes your ally when you know what keeps and when it will be ready to use.

Growing high-value crops is a quiet commitment. It's the promise that tomorrow's meals won't depend on someone else up the supply chain. When the harvest hits the table, you're not just eating—you're proving the kitchen can keep working, even when the world outside slows to a standstill.

SMALL-SPACE FOOD PRODUCTION

A well-organized yard becomes a workshop for food, not a decorative patch. You don't need a farm to feed a family; you need a plan that multiplies the space you've got and makes every inch count. Even in small lots, usable land exists. It just requires clever layouts, vertical growth, and priority planting.

Patios and balconies can become productive plots with container gardening and raised beds. A row of five-gallon buckets along a sunny wall can hold lettuce, radishes, herbs, and small peppers. A small 4x6 raised bed beside the porch can produce greens, onions, and carrots for several months. The trick is to think vertical where possible: trellises for beans and cucumbers along fences or against a sun-facing wall multiply harvests without expanding footprint.

Soil quality matters as much as space. Start with a simple soil mix—half garden soil, half compost, with a touch of coarse sand for drainage if your space is heavy clay. Mulch to conserve moisture, keep weed growth down, and maintain soil warmth. Compost kitchen scraps in a dedicated bin. A good compost system turns waste into soil and adds a quiet, durable source of fertility for the next season.

Water management helps too, though you're not relying on city lines. A small rain barrel, plus a leaky faucet trickle to wash produce, makes a difference. If you can, install a drip line or soaker hoses to minimize water usage and deliver moisture right to the roots. You'll be amazed how much you can grow with a modest water plan and careful scheduling.

In practical terms, plan for two to four spaces that can produce a steady stream of greens, root crops, and leafy vegetables from spring through fall. A three-bed rhythm works well: a fast-turning bed for early greens, a mid-season bed for roots and cabbage, and a late-season bed for beans or fall greens. Use the fence line for climbing crops—peas, beans, or cucumbers—that require light and shelter from wind. Keep a section dedicated to herbs near the kitchen door for quick flavor boosts.

A small space, done well, is enough to build a habit of production. It keeps hands busy, eyes on the soil, and stomachs full with fresh, homegrown flavor. The simple discipline of tending a few crops for weeks at a time becomes a reliable thread through uncertain days.

PLANNING MEALS AROUND WHAT KEEPS

A kitchen built on stored staples and steady production doesn't waste momentum on improvisation out of necessity. It runs on a simple principle: plan meals around what can't fail. Build menus from shelf-stable ingredients and pantry-safe harvests, then fill gaps with what grows well in your yard or garden. This approach yields a calendar of meals that don't depend on a grocery run to stay intact.

Start with an inventory rhythm. Each week, look at what you have on the shelf, what's coming up on expiry, and what's in the garden or pantry. Group items by storage life: long keepers (rice, beans, flour), mid-range (sauces, canned vegetables, dried dairy substitutes), and short-term (fresh herbs, root crops with a shelf life of several weeks). This triage keeps you from overbuying and ensures you use what you have.

Next, design a simple meal cycle. A two-week or four-week rotation works for most families. The cycle should mix meals built from staples—beans and rice, lentil soup with tomato, potato hash with onions—with occasional comfort dishes that boost morale. The emphasis is on variety within a fixed framework: you know you can cook palatable meals from your stored goods, and you minimize waste by planning each day's use of ingredients.

Batch cooking becomes a backbone. Make a pot of beans that lasts several meals; simmer a pot of vegetable soup with pantry vegetables and a drizzle of oil. Cook larger portions of grain dishes to reheat as needed. Use leftovers to create new meals: a bean risotto-style dish with stale rice, a hash from potatoes and cabbage, or a vegetable sauté that carries over to a future lunch. The goal isn't fancy; it's dependable.

Taste matters, even in scarcity. A little salt, a pinch of spice, or a splash of oil can turn a plain bowl into something satisfying. The same pot of lentils can feel meaty with a few carrots and a half-cube of stock. Without a steady supply chain, flavor is your ally—something worth protecting with proper storage and careful meal planning.

Finally, plan for the unexpected. Build a rotating plan that includes a few emergency meals using only what keeps and what needs minimal refrigeration, such as canned goods, dried beans, and grains. If a week is lean, you already know how to stretch each item into multiple meals. If a week shines with new harvests, you'll still pair fresh produce with stored staples to create balance and texture.

A kitchen anchored by stored staples and predictable production is a kitchen that endures. It makes meals possible even when markets stall, prices spike, or deliveries falter. When you can feed your family with a plan rooted in the soil and the shelf, you gain a steady, practical kind of security.

THREE
WATER WITHOUT THE TAP

FINDING RELIABLE WATER SOURCES

Water is life, and in a gridless world it becomes the currency of health. A long outage isn't just dark nights and cold rooms. It's a quiet, creeping danger that starts with drinking water that's unsafe or scarce. The good news is you don't have to place all your bets on one source. A resilient household knows at least two solid options, knows how to move water when pumps die, and treats every drop as fragile and precious.

Start with your own property: a dependable well, a rain barrel, and a plan to haul water if you must. A well is the backbone for many families who live on land. If your well is deep and powered by electricity, you'll want a hand pump, a backup power source, or a gravity-fed cistern. If your well has no hand mechanism, you can still draw water from a shallow source with a hand pump or a bucket-and-pulley system. Shade the picture with a secondary source that isn't tied to electricity—rainwater capture is the most straightforward option to build locally.

Rainwater is a quiet workhorse. A modest roof can yield hundreds of gallons after a single storm if you've set up gutters, screens, and a first-flush filter. A 1,000-square-foot roof catches roughly 600 gallons of water per inch of rain. An ordinary Midwest storm can flood a 1,500-gallon tank in a few hours. Size your storage to your average monthly use, plus a buffer for illness or cleaning. A 55-gallon barrel is a handy starter, but plan for multiple tanks if you're serious about resilience.

Surface water—ponds, streams, or irrigation ditches—shouldn't be your primary drinking supply without treatment. It's a robust source for washing, bulk cleaning, and livestock water, but the risk of contamination is higher. If you're near a reliable stream, you can funnel water into a container during the week and rotate it. Treat it as a secondary source unless you've built a proven filtration-and-boiling routine that you trust.

Hauling water becomes a skill all its own. In droughts or grid failures, you'll be moving water by hand from a neighbor's well, a public tank, or a protected rainwater storage area. A sturdy cart or wheelbarrow and a handful of sturdy jugs are worth their weight in gold. Small, portable containers—five- to seven-gallon jugs—make trips feasible without wearing you out. The goal is to keep a minimum viable supply close to the kitchen door, not off in a distant shed.

When you map sources, you're not stacking hopes on a single punch. You're building a web: your well, rainwater, safe hauled water, and a plan for moving it to the kitchen. In practice, that means labeling every source, testing taste and odor, and knowing how your supply changes with the weather or season. Your well might dry slightly in late summer; your rain barrel will fill during the wet season but sit dry in drought. A resilient household tilts toward redundancy, not charisma.

Concrete steps you can take now:

- Inventory potential sources on your property and within a short walk: well, rain catchment, and a surfaced water option.
- Size storage to your needs, with a 1–2 week buffer as a minimum and a 4–6 week buffer if you're in a remote area.
- Install at least two rain barrels and a simple first-flush device that diverts the initial, dirtier flow away from your storage.
- Build a basic water-haul plan: label containers, assign a hauling route, and schedule regular trips so water doesn't pile up in one storm and then sit unused.
- Keep backups for livestock and garden water separate from drinking water, and never mix them without cleaning and disinfection.

Source reliability isn't magic. It's a routine, tested, and rehearsed set of options you can rely on when pumps stop and pipes freeze.

SAFE STORAGE AT HOME

A day without water is a day without dignity. The moment you store water, you begin a discipline. Fresh water is perishable in practice because containers and lids aren't perfect barriers against contamination, heat, or chemical off-gassing. The goal is clear: keep water safe, visible, and ready.

Choose containers that won't impart taste or toxins. Food-grade plastic, glass, or stainless steel are best. Plastic jugs are light and easy to handle but must be food-grade and free of scratches; glass is inert but heavy and fragile. If you rely on metal containers, ensure they won't corrode or leach metals into your water. Airtight lids, clean interiors, and a dedicated space away from fuel, cleaners, or pesticide—these aren't luxuries. They're the difference between a life-sustaining resource and risk.

How much and how often you rotate matters. A practical target is a two-step stock: enough for drinking and cooking for two weeks, plus a separate reserve for sanitation and cleaning for the same period. For a family of four, that translates to roughly 14–28 gallons for drinking and cooking per week, plus another 14–28 gallons for hygiene and cleaning. In total, you're looking at 1–2 hundred gallons in rotation depending on climate, fuel stores, and the number of people.

Rotation is simple: use the oldest water first, label containers with purchase date, and set a recheck date. If you have room, store water in a cool, shaded area with minimal sunlight. Light accelerates algae growth and tastes. If you must place water near heat, use opaque containers and avoid translucent packaging that lets light through.

Hygiene rules still apply. Never store cleaning chemicals near drinking water. Keep a separate bucket for eyes, hands, and dishes—one for washing, one for rinsing. Cleanliness protects your water from becoming a vector for illness.

Water quality hinges on routine care. Sanitize containers before refilling, inspect seals for wear, and replace brittle containers. If you notice a musty odor or a metallic taste, scrub the interior, re-rack your storage, and consider a fresh water fill.

A practical storage plan for most households looks like this: a primary, gravity-fed source close to the house; a secondary rainwater system with at least 200–300 gallons of storage; and a third backup of hauled water, enough for a week per person at a minimum. The moment you implement a simple plan, you gain confidence.

LOW-TECH PURIFICATION METHODS

Purity isn't optional when the plant treats water as a luxury. You have a toolkit of safe, low-tech options that don't require electricity, expensive gear, or a PhD in chemistry.

Boiling remains the simplest, most reliable method. Bring water to a rolling boil for one minute. At higher elevations, extend that to three minutes. Boiling kills biological pathogens that sitting in a tank won't. It's thirsty work, but it's decisive.

Filtration helps break the chain of contamination before you boil. A ceramic filter can remove bacteria and protozoa, while a simple sand-and-charcoal layer can improve clarity and taste. A basic, two-stage setup works well: coarse sediment removal first, then a finer cartridge or cloth filter. If you're building a DIY filter, let gravity do the work. A slow, gravity-fed flow preserves the filter's effectiveness and minimizes breakthrough of particulates.

Settling is the oldest trick in the book. Let water sit in a clear container for 24 hours. Particles sink to the bottom and can be decanted carefully from the top. It's not a stand-alone method, but it buys you time and improves the performance of filters and boiling.

Chemicals remain a practical option when used correctly. Regular household chlorine bleach can disinfect water that's visibly clear. Add enough bleach to achieve a noticeable odor of chlorine and wait at least 30 minutes before drinking. If the water is cloudy, increase the dose and repeat the waiting period. Iodine is an alternative for sensitive individuals, though it's not ideal for long-term use or for people with thyroid issues.

Never rely on a single method. The best practice is to combine filtration with boiling, or chemical treatment, followed by filtration. You'll reduce the risk of parasites, bacteria, or lingering contaminants that a single method might miss.

Concrete, tested routines beat fancy gear every time. Practice a standard purification sequence on a predictable schedule. Build muscle memory so the steps become second nature when you're tired or pressed for time.

WATER DISCIPLINE FOR DAILY LIFE

A finite water supply forces new habits. You don't expand your options; you protect what you've got. The discipline starts with a clear allocation: decide how much you're willing to use for drinking, cooking, cleaning, and sanitation, and then honor the split.

A practical baseline looks like this: each person drinks one to two quarts per day. Cooking and drinking will take the rest; a modest household may need five to seven gallons per person per day for hygiene, laundry, and household chores. The trick is to keep that total constant while your available supply shifts with weather or demand.

Dishes, laundry, and hygiene should run on basins rather than a constant stream. A two-basin system—one for wash water, one for rinse water—lets you reuse water for hand-washing and dishwashing tasks. You'll stretch the supply by cleaning with less water and by reusing gray water where it won't risk contamination, such as garden irrigation or toilet flushing only where permitted and safe.

Revisit every routine that uses water: brushing teeth, rinsing produce, and washing hands. Keep habits simple and repeatable. A kitchen towel placed by the sink, a small jug of water for quick handwashing, and a dedicated sponge for dirty jobs prevent cross-contamination and waste.

The more you plan, the more you protect. Measure your water use for a week, then trim unnecessary consumption. A disciplined household learns to value every drop and to keep a buffer for moments when life pulls harder.

COMMUNITY, CONTINGENCY, AND WATER SECURITY

No household survives a long outage on its own. Water problems spill into every corner—sanitation, health, and morale. A neighborhood built around shared wells, rainwater, and mutual aid can outlast a city's bureaucracy.

Create a small water council with clearly defined roles: one person tracks sources and quality, another oversees storage and rotation, a third coordinates hauling and transport, and a fourth manages the distribution during peak demand. A rotating schedule keeps people involved and ensures no one is left with a stale supply.

Mutual aid shines when systems fail. A neighbor with a larger rainwater harvest can host a shared storage point, while another with a well can lend hours of pumping without relying on power. Simple agreements—who helps whom, who provides water for children and elders, how much water each household contributes during a week—turn a collection of families into a functioning unit.

Transmission routes are where trust earns its keep. Transparent testing, shared records, and routine audits of water quality create a track record that makes everyone confident in the system. You don't prove reliability with promises; you prove it with data and consistent action.

Water planning isn't a lone project. It's a community practice. When neighbors know each other's routines, the line between personal stocks and shared reserves blurs in a good way. The result isn't just surviving a drought; it's maintaining health, dignity, and a sense of belonging when the grid goes quiet. The water you protect today becomes the trust you count on tomorrow.

FOUR
OFF-GRID COOKING THAT WORKS EVERY DAY

CHOOSING BACKUP COOKING METHODS

A blackout doesn't steal dinner; it reveals what really feeds a household. The kitchen that keeps cooking when the grid dies is built on choices, not gadgets. It rests on a few reliable methods that can be operated without electricity, every day, in any weather. The core idea is simple: don't stack your odds on a single stove. Build a small toolkit of cooking methods you can swap between, season by season, space by space.

Wood-fired cooking has a long track record for daily use in homes that rely on heat, not just heat up. A well-installed wood stove or a sturdy outdoor firebox becomes a workhorse for soups, stews, and frying. The heat is steady if you manage fuel and airflow. The aroma of simmering beans and onions filling a room is not a luxury; it's a signal that your kitchen is still functioning. The downsides are real: smoke, chimney maintenance, and the need for dry, seasoned wood. You'll also need a safe, well-ventilated space and a reliable ash disposal routine. The payoff is durability. Wood is local, portable, and stores well when kept dry. It's the kind of fuel you can count on if lines at the gas station vanish and delivery trucks disappear.

Rocket stoves offer a clever upgrade: high efficiency with minimal fuel. Built from simple materials—cinder blocks, metal tins, and a tight combustion chamber—a rocket stove concentrates heat where you want it. It's cleaner than a primitive open fire, and it heats a kettle or a Dutch oven in a fraction of the time. The catch is set-up: you need a small, stable workspace and a safe way to feed and manage fire. In practice, a family that uses a rocket stove often keeps a spare supply of dry wood for days when wind and moisture slow down the burn. The payoff comes in clean air and quick boil times, which matter when you're trying to feed a family on a cold morning.

Camp stoves and small propane burners fill in the gaps. They're portable, versatile, and fast. They shine for quick breakfasts, coffee, or a dinner that doesn't want to wait for a wood fire to catch. The trade-off is a dependence on canisters or bottled fuel, plus the need for a sheltered, dry place to store cylinders. You should plan for a two-pronged setup: a primary method (often wood or rocket) and a backup burner (camp stove) for days when time is tight or weather keeps you indoors. The principle is reliability. If one method falters, another keeps meals on the table.

Solar ovens bring a different rhythm to cooking. They're quiet, emission-free, and perfect for baking bread, warming soups, or cooking delicate dishes when the sun is strong. They falter in winter skies, after rain, or during late afternoons when the sun sinks low. Still, a solar oven expands your repertoire without consuming fuel. When a clear day arrives, it can take the edge off fuel use and give you a novel tool for roasty, slow-cooked dishes.

The strongest plan is practical: use two or three distinct methods that match your space and routines. A rural home might rely on a wood stove for most meals, a rocket stove for quick day-to-day tasks, and a camp stove for backup during weather shifts. An urban household may lean on a compact wood-fired option indoors, a small camp stove, and a solar oven for weekends. Regardless of setup, organize your kitchen into zones: a wood zone, a stow zone for cylinders, and a solar zone where you can access sunlight without moving furniture.

Finally, practice under normal conditions so the switch between methods feels automatic when the grid goes quiet. Build habit around starting the day with a plan: which method will feed breakfast, which will simmer dinner, where the water lives for quick boiling. The goal is a kitchen that keeps working, not a garden of gadgets that sits idle when power fades. Your backup cooking system should be boring in the best possible way—reliable, repeatable, and familiar to every hand in the house.

FUEL PLANNING AND STORAGE

Fuel isn't a luxury when the grid fails; it's capital that has to be carefully managed. A practical household treats fuel as a critical resource with a clear plan, a safe storage protocol, and a rotation schedule. The aim is to prevent scarcity from becoming chaos. Start with the basics: identify your primary cooking fuels, estimate usage, and then build buffers that cover the longest stretch you realistically face without supply.

Wood is the backbone for many off-grid kitchens. It stores well, if kept dry, and it never asks you for a contract or a data plan. The key questions are volume, dryness, and access. How much wood can you stack safely? How long will it stay dry in your climate? In cold climates, plan for several cords of seasoned hardwood per winter; in milder zones, one or two cords might suffice for a season. The storage rule is simple: keep wood off damp ground, stacked in a dry, ventilated space, with kindling nearby for easy lighting. Rotate stocks so the oldest wood is used first. Label stacks by type and expected burn duration to make quick choices during a busy morning.

Propane and other bottled fuels offer flexibility. If your kitchen layout includes a camp stove or a small grill, you'll likely rely on 20-pound cylinders. Plan for at least two spare cylinders per family for a month of steady cooking, plus a spare or two for seasonal spikes. Store cylinders in a ventilated shed, not inside living spaces, and away from heat and sparks. Regularly inspect hoses and regulators for wear, and replace damaged parts before you need them. Create a simple log or calendar that records fill dates and cylinder swap events. This isn't vanity; it's discipline that prevents last-minute scrambles when a cold front rolls in.

Rotation is everything. Treat all fuels like perishable goods: rotate your stock so older stock is used first, and keep a running mental map of what's in the cabinet, what's in the shed, and what's on the shelf. Establish a predictable cadence: check stocks weekly, light a stove or stove-like device weekly, and refuel before you're forced to. If you live near seasonal changes, adjust your buffer sizes in anticipation of harsher months and shorter daylight. The most resilient households don't guess their needs in a crisis; they've already calculated them and kept the numbers visible at a glance.

Above all, balance is the goal. Have at least two fuel options you can access without power, and keep the storage safe and accessible. Adequate fuel planning means you can cook, heat the kitchen, and stay sane even when delivery trucks don't roll.

HAND TOOLS FOR FOOD PREP

Power abhors a vacuum, but it should not dictate every kitchen task. Hand tools liberate your cooking from electricity and force you to rethink routines. A simple, robust toolkit makes grain work, vegetable prep, and meat handling predictable and repeatable, even in hardship. The right tools reduce effort, increase accuracy, and cut waste.

A sturdy grain mill is a frontline tool for true food security. Manual mills crank out flour from wheat, corn, or other grains with a steady rhythm. Look for a design with a solid screw or burr mechanism and a body that won't twist under pressure. A mill is worth it if you bake regularly or you're saving seed for future sowing. It won't replace power, but it makes every seed worth the work. Pair it with a good supply of dried grains and a simple, reliable recipe for bread and porridge.

Peelers, paring knives, and kitchen shears are the everyday backbone. A good Y-peeler handles carrots, potatoes, and apples with clean, thin peels. A sharp paring knife handles delicate work—garlic skins, citrus segments, and herb trimming. Shears cut herbs, poultry cords, and stringy greens without wincing. Keep a sharpening stone near the workstation and schedule a quick touch-up after each big task. A dull tool is a wasted breath.

Choppers and slicers offer speed without electricity. A hand-cranked slicer is perfect for apples, cabbage, and onions when a food processor isn't available. A simple meat cleaver or chef's knife does heavy lifting for chopping root vegetables and herbs. Modern push-style choppers can be convenient, but they rely on clean, dry blades to stay safe and effective. Practice safe cutting habits: keep fingertips tucked, work on stable boards, and maintain a clean, dry surface to prevent slips.

Dough, dough boards, and a rolling pin round out the core kit. A wooden board and a heavy rolling pin let you make bread, tortillas, and flatbreads by feel, not by electronics. Cast-iron skillets and Dutch ovens complete the set. They heat evenly and retain heat, which matters when you're cooking for several hours in an indoor oven or over embers. Keep cast iron clean and seasoned; it lasts generations and doesn't demand a power supply to stay useful.

Maintenance matters as much as the tools themselves. Oil hinges, sharpen edges, and keep blades dry to prevent rust. Store tools in a dedicated, organized space so you can grab them by habit rather than rummaging. A routine of sharpening, cleaning, and replacing worn parts is a quiet, powerful form of preparedness.

Hand tools pay back in spades. They let a family cook with what's on hand, stretch groceries, and keep dignity intact when electronics go dark. The goal isn't nostalgia; it's reliability under pressure.

COOKING SIMPLE STAPLE-BASED MEALS

A kitchen that sustains a family without electricity thrives on a small repertoire of staple-based meals. The blueprint is simple: build meals from flour, beans, grains, root vegetables, and canned goods, then adapt with what you harvest, preserve, or barter. The power comes from structure, not novelty. A repeatable approach to cooking gives every family member a defined role, builds confidence, and reduces waste.

Start with a pantry that favors bulk staples: dried beans, lentils, rice, oats, cornmeal, flour, canned tomatoes, dried milk, salt, sugar, and a handful of dried herbs. These are the fuel that feeds you when the world is crowded with supply-chain noise. Each week, set a plan that rotates these staples into meals that share components—beans and rice, cornmeal mush, or oat porridge—so you're not reinventing the wheel daily.

One-pot meals are your friends. A large pot of bean soup, simmered with onion, garlic, and a root vegetable, feeds a crowd and freezes well. Add a handful of rice or barley to stretch it, and you've created a nourishing, low-energy dish. A simple lentil stew with carrots and potatoes becomes heartier when you finish it with a splash of tomato and a pinch of dried thyme. Both lean on the same family of ingredients and cooking methods, making a busy day predictable.

Batch cooking matters. On a morning with sun and time, prepare a big pot of beans or a hearty soup. Use the leftovers for lunches, reheat with a few spices, and pair with fresh bread or flatbread. A pot of porridge in the morning, thickened with a spoon of honey or dried fruit, keeps energy steady for several hours. A cornmeal mush hot from the Dutch oven makes a comforting base for vegetables and a fried egg, if you've got them.

Root vegetables deserve their own spotlight. Potatoes, beets, carrots, and squashes store well in a cool, dry corner. They can be roasted in a pan or added to soups and stews. Keep an eye on aging produce and use up soft vegetables first in a sturdy veggie fry or a hearty hash. A simple cabbage and potato dish, fried with onions in a cast-iron skillet, turns into a filling meal with the right seasoning.

Flavor comes from salt, herbs, dried tomato, and a touch of fat or dairy when available. A pot of beans tastes richer if you finish with a drizzle of olive oil or a spoon of rendered fat. A loaf of bread—lean dough, slow fermentation, hot pan—helps stretch every bit of grain. The core meals aren't fancy; they're dependable, repeatable, and adaptable to what's in the pantry.

The disciplined routine is the secret. Decide on two core meals and build a week's plan around them, with a few variations to use up leftovers. Track the results: which meals pair best with your storage, which cook times feel realistic on a busy day, and which ingredients you're constantly over- or under-buying. Your goal is not gourmet, but dependable, nourishing, and flexible enough to weather uncertainty. A household that treats staples as a function, not a fad, cooks through weeks with calm and dignity.

FIVE
HEAT, LIGHT, AND SHELTER IN A COLD HOUSE

HEATING ONE SPACE WELL

Heat concentrated in one room changes a crisis into endurance. A warm core keeps bodies warmer, keeps pipes from freezing, and steadies nerves when the grid goes quiet. The goal isn't luxury; it's clarity. Pick a room you use most—the living area or the kitchen—and make it your heat hub. Its warmth will radiate through the rest of the house in practical, survivable ways. Start with a plan, not a gadget pile, and you'll see a room become a shelter rather than a trap.

Choose the hub room for its layout and access. It should be interior or partially shielded from drafts, with a door to separate it from the cold zones. A south- or west-facing wall helps gather sunlight during the day, even in winter, and a masonry or brick hearth nearby acts as a natural heat store. If there's a central chimney or a single pipe run, position the main heat source nearby so the warmth can circulate. A stove or a safe, purpose-built heater becomes the room's heartbeat. Clearance matters. Keep combustibles at least three feet away, use a noncombustible backing on the wall behind the stove, and install a sturdy hearth protector if you're unsure about floor protection.

The heat source should anchor a simple circulation plan, not a gadget obsession. A wood stove, pellet stove, or propane heater can work, but only if used with proper air exchange and ventilation. In a cold house, you don't want to overfire. You want steady warmth that you can sustain with modest fuel. Set a routine: a load of wood in the morning, a soft simmer of heat through the afternoon, a slower burn overnight. Consistency beats bursts of warmth that burn fuel and exhaust energy.

Distribute heat without wasting it. Close doors to unheated rooms to prevent heat from leaking away. Put a sofa or an armchair around the heater to create a small heat chair—where people gather, heat stays longer. If your room lacks built-in mass like stone, create it with a heavy rug, brick or tile on the floor, or even a sturdy wooden bench with a thick cushion. Heat loves weight; the room becomes warmer when you provide a surface that soaks and radiates that energy after the fire dies down.

Ventilation is not a luxury here; it's a requirement. A well-ventilated space is safer and more comfortable than a stale one. A small vent or window cracked open by a person at the stove helps air exchange without losing heat to the outdoors. A ceiling or box fan set on low helps move warm air that climbs into the ceiling back down to floor level. Don't chase the idea of perfect insulation on the first week; start with controlled heat, then tighten the envelope around it.

Bedding and clothing also matter. Heavier quilts, flannel sheets, and wool blankets trap heat near bodies as temperatures drop. Layering means warmth without burning through fuel. The eye of your home becomes the bed area; with careful habits, you'll sleep with confidence instead of fear. In this setup, you're storing warmth in practical ways—heat in the room, warmth in the body, and routine in the day.

A concrete example: a two-story, drafty farmhouse with a wood stove on the first floor. The family built a heat hub in the front room, installed a heat shield, and added a thick rug and heavy curtains. They closed off bedrooms except for a door to this hub, used a ceiling fan on low to move air, and planned two burns per day—morning for warmth, evening to keep the chill at bay. That single, steady warmth changed the mood of the house; fear of the cold faded as people settled into a predictable rhythm. The warm room isn't a luxury in a blackout; it's the anchor that keeps the whole home functional.

The bottom line: a warm core makes everything else possible. When the house stabilizes around one inviting space, cooking, chores, and even care-giving flow more smoothly. The climate inside the room reflects how well you've controlled the climate outside it. Keep the hub clean, well-sealed, and well-lit with safe, non-electric sources, and you'll give your household a fighting chance when the lights stay off.

WOOD, COAL, AND BACKUP FUEL CHOICES

Fuel is not a decorative choice; it's a critical element of daily life during an outage. The decision comes down to density, accessibility, cost, and safety. You'll sleep better if you know what you're burning, where it's stored, and how you'll move it when the snow piles up. The Amish-inspired approach is simple: a few reliable options, stored properly, and rotated with the season.

Wood remains the most versatile heat source for a long outage. It's renewable, can be produced locally, and does well in a properly sized stove or fireplace. The key is season and dry storage. Dry wood is lifesaving: moisture below 20 percent is a good target. Split and stack to shed moisture quickly, protected from rain and ground dampness. A cord of seasoned hardwood can feel pricey, but it's a steady supply that you can manage with a few hours of winter work. If you have limited space, consider a single, well-placed wood stove or a heat-focused fireplace with clean flues and a heat shield. Wood heat rewards discipline: you'll burn in the morning, maintain through the day, and reduce the number of reloads at night.

Coal offers high energy density and long burn times, which can reduce fuel handling in the depths of winter. It's less forgiving than wood—dust, soot, and heavier clinker management demand care. If you choose coal, store it in a dry, ventilated bin away from ignition sources. You'll need a capable ash handling system and a stove or heater rated for coal, with a proper grate and ash pan. Coal heat can be steady, but you'll pay in dust control, time, and a more involved cleaning routine.

Propane and kerosene bring convenience. A propane space heater or portable tank gives quick warmth and predictable performance, especially in milder climates or in rooms with poor wood-burning ventilation. Propane storage requires outdoor, shaded placement, with cylinders kept upright and protected from sun. Kerosene lanterns and heaters provide reliable light and heat, but they demand careful handling, clean ventilation, and dedicated fuel storage held away from living space. If you lean on these fuels, establish a rotating stock—don't let gasoline or kerosene sit so long it becomes unusable.

Backup fuel strategies are not about abundance but reliability. Use a small, clearly marked fuel shed with shelves for each type, labeled containers, and a simple rotation wheel. Check moisture levels, test stoves, and replace seals when necessary. A practical household keeps a simple ledger: what's stored, where it is, and how long it will last for the household in a worst-case week. The goal is to avoid gaps that force sudden, stressful hunts for fuel.

Diversity matters. A mix of heat options reduces risk: a wood stove paired with a propane heater for high-demand windows, a coal option for cold snaps, and a kerosene lamp or lantern for light without electricity. It's not elegant, but it's robust. You'll have more control over your comfort when you know you're not dependent on a single fuel stream. The trick is to create a predictable pattern of use and storage that your family can follow without a sermon. It's about making the right choice, then executing it calmly and consistently.

DRAFT CONTROL AND INSULATION UPGRADES

A draft may be invisible, but its effect is loud and immediate: cold air comes in, warm air slips out, and the room never feels truly comfortable. The first step is to stop leaks; the second is to trap heat that's already there. Start with the simplest, most cost-effective fixes. You don't need a complete house remodel to make a real difference. Small investments in insulation and sealing compound accumulate into real warmth over a season.

Seal the big leaks first. Doors and windows are the usual suspects. Install door sweeps on exterior doors; add weatherstripping around frames; focus on the bottom gaps where cold air hisses in. Old windows can be drafty; add clear plastic sheeting or thermal curtains to cut pine-scented drafts during the winter months. If you don't have room for storm windows, use heavy, lined curtains that close tightly at night. The effect is measurable: rooms feel warmer because cold currents won't rush through every crack.

Insulation can be improved without major renovation. In attics, add insulation where there's room rather than replacing the whole ceiling. If you can't increase attic thickness, seal the hatch with a tight-fitting cover and weatherstripping. On floors, a thick rug or a pair of layered runners reduce heat loss through bare boards and concrete. In a room with a stove, a heat shield behind the stove helps protect walls and adds a little more efficiency—heat absorbs into the shield and is slowly radiated back into the room.

Draft-proofing is not glamorous but it is effective. Draft snakes on doors, sealed electrical outlets, and anchoring lightweight curtains to walls can stop convection loops that drain warmth. The aim is to slow the wind tunnel in your walls, not to create a hermetic box. You want steady warmth with airflow that's controlled, not a room that feels like a refrigerator with a burn barrel in the middle.

Mass helps. Place a few heavy, heat-retaining objects near the stove. A brick or stone bench, a clay pot filled with water, or ceramic tiles placed along a wall near the heat source can store warmth when the fire is low. In a pinch, you can set pails of water or bricks near the heater. Heat is absorbed and released slowly, smoothing the peaks and keeping the air more uniform. None of these moves requires a contractor; they're practical, repeatable, and easy to maintain.

Finally, the layout matters. A compact floor plan makes it easier to trap heat. If possible, keep sleeping areas adjacent to the heat hub, rather than on exposed exterior walls. The result is a house that feels smaller and easier to manage when the lights go out. The goal is to minimize cold air exchange while maximizing the warmth you've built up during the day. A little planning yields a home that stays comfortable with less fuel and less effort.

LAMPS, LANTERNS, AND FIRE SAFETY

Lighting is not an afterthought when the power fails; it's a fundamental part of daily life. Non-electric lighting must be reliable, safe, and easy to manage. Oil lamps, kerosene lanterns, and hand-cranked flashlights all have a place in a well-prepared home. Each option brings a different rhythm to the evening routine, and each requires discipline in storage, handling, and placement. The right mix lets you illuminate working spaces for chores, highlight reading corners for kids, and keep hallways visible for safe movement.

Choose lamps that fit your space and your fuel. A few properly chosen hurricane lanterns or lanterns with glass globes provide bright, consistent light and are easier to protect from drafts. Oil lamps require clean wicks and steady trimming. Keep spare mantles and wicks on hand, along with a small toolkit for quick repairs. For nights with extra activity, place lighting at key work zones: the kitchen counter for food prep, the workshop bench for repairs, and the entry for safe movement. Avoid clustering many light sources around the same small area, which can overheat surfaces and create glare that tires eyes.

Safety matters more than brightness. Never burn a lamp in an unventilated space for long; smoke and fumes accumulate, and the risk of fire increases. Provide adequate ventilation by cracking a window slightly or using a small vent. Keep fuels in sturdy metal containers away from ovens, stoves, and possible ignition sources. Clear a safe zone around every lamp—no drapes, no paper, no kindling within reach. Place lamps on stable surfaces, preferably metal trays that can catch a spill or a knocked lamp. Make sure kids understand not to touch hot glass and always to ask before moving anything that's lit.

A simple rule anchors safety: treat every lamp as a small fire in a glass bottle. Don't rush. Check the fuel level each day, rotate stock to keep it fresh, and replace worn components before they fail. Fire safety in a non-electric home isn't dramatic; it's about steady, informed habits. A single well-lit room, a handful of reliable lamps, and a clear plan for what to do if a lantern flickers keep danger at bay and morale intact when the grid stays quiet.

SIX
SANITATION, LAUNDRY, AND THE DIRTY WORK OF STAYING HEALTHY

When the taps vanish, a clean bathroom isn't a luxury—it's a line of defense against illness. In a long outage, waste management becomes as essential as heat and water. Without a functioning sewer system or municipal trash pickup, germs ride on the air and water, and odor becomes a health risk. This chapter treats sanitation as a set of repeatable routines, not a one-off fix. The goal is to keep waste contained, water-borne disease at bay, and family morale intact by turning a basic human need into organized practice.

EMERGENCY TOILET SYSTEMS: BUCKET, COMPOSTING, AND SAFE WASTE HANDLING

A proper toilet strategy is a chain of predictable steps, not a guesswork compromise. In many off-grid households, the simplest tool is a sturdy five-gallon bucket with a lid. A basic setup uses a lined bucket, a secure lid, and a carbon layer.

Sawdust, wood shavings, shredded paper, or peat moss absorb moisture and mask odor. A common rhythm is to designate two or three buckets in rotation: one in use, one drying, one for waste awaiting disposal. A small, tight-fitting seat that fits over the bucket turns it into a usable toilet while you work out placement and disposal.

Site selection matters as much as the device itself. Choose a spot at least fifty yards from any well, spring, or vegetable garden, and uphill from surfaces that could carry water toward the house. If a trench or pit is available, line it with a durable membrane or, more simply, bury waste in a shallow pit and cover with soil after each use. In cold weather, a double layer of bedding helps keep odors down and moisture under control.

A composting toilet is a more elaborate option that pays back over time. It relies on two chambers: one for recently deposited waste (which also receives carbon bedding) and one for aging material. Urine and solid waste are managed separately; urine can be diluted and applied to non-edible plants or stored for later use if you have a suitable layout. The key is carbon balance and airflow. Add carbon-rich material with every use—the more bedding, the less odor and the faster moisture evaporates. Keep the chamber moist enough to allow microbial activity but not so wet that it clumps and becomes anaerobic.

Disposal rules prevent waste from becoming a health hazard. If you don't have a dedicated compost site, bury waste at least a foot deep and away from vegetable beds and water sources. Do not mix toilet waste with kitchen scraps that attract rats. If you must store waste briefly, seal it in a sturdy bag or bucket with a lid and keep it in a sheltered, ventilated place away from living areas. Never burn waste that may release toxic fumes; instead, manage it with containment and time.

Hygiene after use is non-negotiable. Wash your hands with soap and water whenever possible; when water is scarce, wash with a small amount of soapy water and dry with a clean cloth, then wash again when more water is available. Clean the toilet area with a simple solution of soap and water, or vinegar if you're avoiding stronger chemicals. Odor control hinges on lids, aeration, and regular removal of full buckets before they overflow their capacity. With clear routines and proper containment, waste no longer undermines health or comfort.

These toilet strategies are not a relic but a practical toolkit for a future where infrastructure can't be relied on. A well-planned system reduces illness, preserves dignity, and keeps your home moving forward even when the grid is quiet.

LAUNDRY AND CLOTHES WASHING: LOW-WATER METHODS AND DRYING ROUTINES

Clean garments are a shield against infection and fatigue alike. In a grid-down scenario, laundry becomes a disciplined chore, not a weekend luxury. A traditional hand-wash station uses two sturdy basins or a wash tub, a washboard for stubborn stains, and a hand-cranked or lever-wringer for squeezing out water. A practical setup might include a third basin for rinsing and a rack or line nearby for drying. The goal is to minimize water use while maximizing cleanliness and fabric longevity.

Soaping happens in the soap basin. A simple bar soap or homemade lye soap works well for fabric care without relying on modern detergents. Pre-treat stains with a paste of soap and a little water or rub heavily soiled areas with a bit of baking soda. Work from the cleanest areas to the dirtiest to avoid re-depositing grime. Rinse by dunking the fabric into the rinse basin and wringing out as much water as you can. Reuse rinse water for other tasks that won't transfer dirt to your body or food, such as dishwashing or tool cleaning, then drain into gray-water-safe outlets or garden soil when appropriate.

Wringing is a mercy in thin times. A manual wringer attachment lets you squeeze more water from fabric with less effort, reducing drying time. If a wringer isn't available, twist the fabric between your hands and then squeeze along the edge to extract as much water as possible. Drying is fastest when clothes are hung on a line or rack in a sunny, breezy spot. In cool or damp weather, lay garments on a clean towel to wick moisture, then finish drying on a rack indoors near a heat source, such as a wood stove or solar-heated surface.

Separate the big jobs from the small. Heavily soiled work clothes deserve extra attention and possibly a second wash, while everyday wear can share rinse water from the main load. Keep a dedicated bin for underwear and socks to reduce cross-contamination, and wash them more frequently. Storage of clean clothes should be dry, on breathable hangers or folds that minimize creases and mildew risk. A family four can go several days between full washings with smart batching: rotate daily tasks, pre-treat stains, and use sun and wind to your advantage.

An old gardener's habit offers a smart trick: compost the water you've poured onto the soil after it's cooled. It returns moisture to the earth and reduces waste, keeping your yard productive while reducing the drain on your water reserves. Laundry isn't glamorous, but it's a reliable routine that limits grime, preserves fabrics, and protects skin from irritation when soap and water are scarce.

BATHING WITHOUT RUNNING WATER: PRACTICAL HYGIENE ROUTINES

A bath can be done with a kettle, a basin, and a plan. Daily hygiene carries a heavy load in a crisis, not just for cleanliness but for morale and health. A full bath every few days can be sustainable when water is tight; in between, you can perform efficient sponge baths over a basin and a cloth. Start with warm water heated on a wood stove, then pour into a basin to create a bathable temperature. Use the same soap as for laundry to reduce chemical variety and costs.

Wash from the face down, then arms, torso, and legs. If you're rinsing, use a smaller amount of water and pour gently to minimize cold shock. Hair can be washed with a modest amount of water, focusing on the scalp and roots. For many households, hair washing is a weekly or biweekly routine; on days between, you can use a dry shampoo or wipe with a damp cloth plus powder on the scalp to reduce oils. After bathing, dry quickly with a clean towel and dress in dry clothing to avoid chill.

Storing and drawing water for bathing require thoughtful planning. Keep water barrels or drums clean and capped, with a first-in, first-out rotation to prevent stagnation. If you have hot water for washing dishes or cooking, consider using some of that water for body washing when needed. Gray-water rules apply again here: avoid dumping bath water near edible plants or into streams that people use for drinking or washing.

When possible, incorporate warmth and light to elevate mood during cold weather. A small, safe lamp can provide enough heat to evaporate damp air, reducing mildew in bathroom corners. A well-worn wooden stool and a stable, slatted rack help keep towels dry and accessible. The goal is simple: stay clean enough to stay healthy, using only the water and fuel you already have.

KEEPING PESTS AND DISEASE DOWN: FLY CONTROL, SURFACE CLEANING, AND HYGIENE RULES

A clean home is a shield against illness and pests that thrive on neglect. Flies thrive where waste and food residues sit too long. Start with waste handling as a preventive measure: keep toilet areas sealed, bagged, or composted, and cover garbage with a tight lid. Regularly remove full waste containers and store waste away from living spaces. Husbanding pet waste is part of this, too. A simple rule is to treat any waste as a health hazard until it is out of reach and contained.

Daily cleaning streams into larger routines. Wipe down countertops, tables, and high-touch surfaces with soap and water. If you want a stronger effect, a simple solution of water and white vinegar works well for many surfaces and is vastly cheaper than commercial cleaners. For tougher stains, a paste of baking soda and water can lift grime without harming fabrics or finishes. Pay particular attention to cutting boards, sinks, and floors near food prep areas, because contamination travels quickly.

Insects and waterborne disease share a habit: they ride along with neglect. Prevention starts with keeping food sealed and surfaces dry. Use screens on doors or a simple flapping door curtain to reduce entry points for insects. When flies appear, set simple homemade traps using sugar water in a small jar with a paper funnel. Redirect moisture away from living spaces by arranging drainage that carries it toward gardens rather than into basements or kitchens.

Hygiene rules become daily practice: wash hands after handling waste, after changing dressings or bandages, after touching animals, and before meals or food prep. People, especially children, should avoid putting hands in their mouths after handling waste or touching animals. Shared spaces benefit from a weekly deep clean, with particular attention to floors and corners where dust and fecal particulates can accumulate.

Mutual aid plays a role here, too. Neighbors can rotate duties —someone handles the bucket systems while another tends the laundry line and another oversees pest control. A small, trusted network makes the work lighter and more reliable over time. Sanitation isn't glamorous, but it creates a stable home where the family can eat, sleep, and work with confidence even when the grid is quiet.

Overall, sanitation, laundry, and daily bathing are not just chores. They are ongoing health work—reliable systems that keep illness from spiraling when water, sewer, and power are scarce. With clear routines, durable tools, and shared effort, a household stays clean, efficient, and capable in difficult times.

SEVEN
BASIC MEDICAL CARE WHEN HELP IS FAR AWAY

BUILDING A SERIOUS HOME MEDICAL KIT

A medical kit is the quiet engine of resilience when the grid stays dark and doors stay shut. It isn't a gadget stash or a gadget chest. It's a carefully chosen, regularly refreshed set of tools that keeps a family moving through common injuries, fevers, and ordinary illness when a trip to the clinic isn't possible. Start with the basics, then build in redundancy. The goal is readiness, not gadgetry or guesswork.

Stock the essentials in two places: a primary kit in the kitchen or pantry and a smaller, clearly labeled subset in the car or workshop. The kitchen kit covers most daytime needs; the secondary kit catches you on the road or during chores. Keep the kits in a labeled box or a durable case, protected from heat and moisture and out of reach of small children. Rotate stock every year. When items reach their expiry, replace them. A stale tube of antibiotic ointment is worse than no ointment at all, and an old bottle of antiseptic is not more effective, just unsafe longer.

The core components fall into five categories: wound care, medications and fever relief, hydration and gut care, tools and PPE, and documentation. In each category, quality matters more than quantity. A clean, properly sealed gauze pad is more useful than a stack of half-used bandages. A dependable digital thermometer is better than a dozen unreliable ones. Better to have fewer, well-chosen items that you know how to use than a warehouse of stuff you ignore.

Wound care begins with clean, accessible supplies. Pack sterile gauze pads, assorted bandages, adhesive tape, sterile wipes, and a small bottle of antiseptic solution. Include sterile saline for wound irrigation, a pair of trauma shears, and non-adherent dressings so you don't pull off healing tissue when you remove bandages. Butterfly closures and a few square inches of medical tape can close small cuts when stitching isn't an option. Have disposable gloves, a compact first-aid manual, and a few alcohol-free cleansing wipes. Keep a mini-cooler and a few cold packs for swelling. Keep a bottle of burn-specific gel or a clean, cool cloth for burn soothing; avoid applying greasy ointments to serious burns.

For fever and pain relief, pick OTC medications that cover both adults and children. Acetaminophen helps with fever and mild pain; ibuprofen reduces inflammation and also eases pain. If you have children, stock appropriate children's formulations and a dosing guide. Include electrolyte packets or oral rehydration solution for dehydration from fever, vomiting, or diarrhea. A small supply of antiemetic meds—

again, age-appropriate—can save hours when someone feels overwhelmed by nausea.

Hydration and gut care are often overlooked. A simple oral rehydration solution can keep a person from slipping into dehydration after a stomach bug or a hot day. Keep bottled water, a sealed pitcher, and clean cups for family use. A basic set of electrolyte sachets, plus a bottle of electrolyte solution you can mix with water, is worth having. A few doses of a gentle antidiarrheal can help, but use only as directed and be mindful of children's needs. If a household member is pregnant, elderly, or immunocompromised, adapt gut-care supplies to those risks.

Tools and personal protective equipment round out the kit. Nitrile gloves, a small flashlight, extra batteries, a thermometer, a nasal aspirator for kids, a pair of tweezers, and small scissors are practical in daily life and emergencies. A basic CPR-face shield or mask is worth having if someone in the home is trained in CPR. A knot of rope and a few triangular bandages can serve as improvised splints or slings. Label everything clearly and keep a simple, laminated list of contents in the lid of each kit so a non-specialist can locate what they need in a hurry.

Finally, documentation turns a scattered set of items into a coherent plan. Keep a current list of all household medications, dosages, allergies, and chronic conditions. Include doctor contact information, nearby clinics, and the locations of the nearest hospital. Store copies of prescriptions and immunization records, plus a list of known adverse drug reactions. Keep a small notebook or laminated sheet noting any changes in health, new symptoms, or reactions to meds.

A well-maintained medical log helps you spot patterns and prevents duplicate or unsafe dosing.

A practical approach is to inventory your needs twice a year. One spring check for outdoor or seasonal risks; another in late fall to prepare for cold weather. When you finish a season, audit what you used, what you avoided, and what you need to restock. A sturdy kit built with disciplined, repeatable routines becomes an invisible backbone when the power is out and the store shelves sit bare.

FIRST-AID SKILLS WORTH PRACTICING

A toolkit without know-how sits idle. Skills turn equipment into outcomes. The simplest, most universally useful abilities belong to everyone: controlling bleeding, cleaning wounds, recognizing fever, and knowing when a problem exceeds home capacity. Practice makes the difference between a calm response and a panicked stumble.

Bleeding control is first on the list. Direct, steady pressure with a clean cloth or gauze slows most bleeds fast. If blood soaks through, apply a second layer without removing the first. Elevate the wound if possible and continue pressure. A compressive bandage over a larger area can manage heavier bleeding. Avoid removing dressings to "check the wound" during the first minutes; if the bleeding won't stop, consider a more secure wrap or a tourniquet as a last resort after calling for help. In a family where there's a trained adult, teach everyone how to apply pressure and know when to escalate care.

Wounds should be cleaned as soon as bleeding is controlled. Rinse with clean water, removing dirt or debris with sterile tweezers if you can do so without causing more harm. Do not scrub a wound aggressively. Pat dry and apply a thin layer of antibiotic ointment if it's available; cover with a sterile dressing. Change dressings daily or when they become wet or dirty. If the wound shows redness, swelling, warmth, or pus after 24 hours, seek medical advice.

Burns require different care. No ice. Cool running water for 10–20 minutes, then loosely cover with a clean, non-fluffy cloth or non-adhesive dressing. Do not apply ointments to open burns, and avoid popping blisters. Large or deep burns, burns on the face or hands, or burns surrounding the joints demand outside care as soon as it's available. In a long outage, the best wound is a clean wound; second best is a covered wound tailored to the person's size and needs.

Fractures and sprains need immobilization. If you suspect a break, keep the limb still and supported. Use a rigid splint—wood, cardboard, or a sturdy stick—padding as needed. Do not attempt to realign a bone. Elevate the injured part above heart level when possible and apply a cold pack to limit swelling. Severe pain, deformity, numbness, or an inability to move the limb are signs to seek professional help—even if that means traveling to a distant clinic.

Fever and dehydration demand steady attention. Offer small sips of water or an oral rehydration solution and keep the person comfortable. If a fever remains high for more than two days in an adult, or if a child has a fever lasting longer than 24 hours or shows poor intake, lethargy, or confusion, call for outside help. Fever can be a simple virus, but it can also mask something more serious. When in doubt, err on the side of caution.

Infections sometimes look minor but can worsen quickly. Cleanse and cover, monitor for increasing redness, warmth, or drainage, and watch for fever or swelling that spreads. A course of topical antibiotic can help routine minor infections if you have a doctor's guidance, but don't rely on antibiotics as a routine solution. If an infection seems to spread or you're unsure about a wound's depth, seek outside care promptly.

A quick word on CPR and airway emergencies. If someone stops breathing and you're trained, perform CPR according to the latest guidance you've practiced. If you're not trained, call for help and assist with basic life support if you can do so safely. Keep a space where you can practice these skills with a partner—this isn't about memorization; it's about muscle memory.

Practical drills pay off. Schedule a quarterly practice session for family members to run through bleeding control, wound cleaning, and burn care. Rotate roles so kids understand what to do when an adult is busy in the kitchen, or when a caregiver is ill. Simple drills normalize these responses and reduce anxiety when real trouble arrives.

MANAGING CHRONIC CONDITIONS

In emergencies, chronic conditions don't pause. They demand planning, consistency, and predictable routines. The backbone is a supply strategy that acknowledges limits: pharmacies will be closed, supplies will be inconsistent, and access to clinicians may be delayed. Your plan should be proactive, not reactive.

Begin with an honest inventory of everyone's health needs. List every prescription, its dosage, how often it's taken, and the preferred time of day. Note any allergies or special requirements, such as refrigeration. Attach copies to your medical kit and keep digital records on a non-networked device if possible. The key is to know exactly what is on hand and what isn't—before you run out.

Medications require careful handling. For chronic conditions, keep a surplus of necessary medicines when possible, and extend renewals with your clinician when you can. If a pharmacy is unlikely to reopen quickly, plan ahead for a longer supply. This is not about hoarding; it's about continuity of care. Store medications in a cool, dry place away from sunlight, in child-resistant containers when feasible. Rotate stock with every season change and discard anything that's clearly labeled unsafe.

Diabetes care changes in a grid-down scenario. Maintain a steady routine for blood sugar checks if you have a meter; stock extra test strips and batteries. Keep a small supply of fast-acting carbohydrates, such as glucose tablets or candy, for hypoglycemia. Insulin requires careful handling; if refrigeration becomes unreliable, consult a clinician about a safe extension plan and practical storage. If monitoring tools fail, rely on how you feel, act on symptoms, and have a clear plan for seeking help.

Hypertension and heart risk don't disappear when the lights go out. Keep medications available, monitor your own symptoms, and adjust activity and salt intake to the limits of your space and climate. If chest pain, severe shortness of breath, or fainting occurs, treat it as an emergency and seek outside help immediately. If access to care is limited, keep a written record of blood pressure readings and any symptoms to discuss with a clinician when it's possible.

For respiratory conditions, asthma, COPD, or allergies, ensure inhalers and spacers are labeled and easy to reach. Practice proper inhaler technique with family members so that, in a crisis, every heart and mind knows how to help. When air quality deteriorates in a power outage, plan for safer indoor ventilation, minimal outdoor exposure, and a quick route to medical aid if symptoms worsen.

Mobility challenges require anticipatory planning. If daily routines rely on equipment like walkers or canes, have spares or simple repairs, and practice transfers and safe movement with a partner. In a grid-down world, independence hinges on how well you adapt supports, not on the tools themselves.

Your medical routine should be visible, repeatable, and embedded in daily life. Post a simple schedule in a central location: medication times, refill reminders, and symptom check-ins. Use a shared calendar or a laminated sheet to keep everyone on the same page. If you have children, involve them in the process: teach them how to measure, count, and observe. The objective isn't perfection; it's continuity—keeping people healthy with steady, predictable habits.

KNOWING WHEN TO SEEK OUTSIDE HELP

Some problems outgrow home care, especially where chronic conditions intersect with aging, injury, or infection. The decision to seek outside help is a disciplined choice, not a hurry, and it rests on clear, observable signals rather than fear.

Red flags are concrete. Chest pain, fainting, or new confusion in an adult signals a possible cardiac or neurological crisis. Heavy or uncontrolled bleeding, numbness or weakness on one side of the body, sudden severe headache, or slurred speech require immediate professional assessment. A fever with stiff neck, dehydration signs in a child, vomiting that won't stop, or a deep wound showing tissue or bone should trigger a call or an on-site visit as soon as possible. If there's any doubt, treat the situation as urgent until proven otherwise.

When to call local help depends on the time to reach care. If you can reach a clinic within an hour, make the call and arrange transport. If transport is delayed, arrange a neighbor or family member to assist. In a remote area, keep a list of the closest facilities, their hours, and any after-hours number you can reach through non-digital means. A printed phone list on the wall reduces delays when phones are down or dead batteries rule out smartphones.

Triage isn't guessing. It's a rational assessment. If symptoms could be managed safely at home with a short window for observation, you document the care given and wait. If symptoms worsen or new warning signs appear, escalate promptly. You do not wait out the clock on a condition that could deteriorate rapidly.

If you do call for outside help, arrive with a plan. Bring a written summary of symptoms, medications, allergies, and recent changes in health. Have a list of recent vital signs if you track them at home. When professionals come in with limited time and resources, you've already done the groundwork—so you can get the care you need faster.

TRAINING, RECORDS, AND ROUTINE

Consistency matters more than grand gestures. Build a simple, repeatable medical rhythm that fits your everyday life. The idea is not to become a clinic, but to stay functional and calm when demand spikes.

Create practical routines around two pillars: routine replacement and skill refresh. Routine replacement means stocking the kit, rotating meds, and rechecking expiration dates on a set schedule—every six months for some items, yearly for others. Skill refresh means quarterly drills, a quick family review of steps for bleeding control, wound cleaning, and fever management, and a rotation where each family member takes a turn leading a mini-scenario.

A small, neutral checklist can anchor your routines. Keep it near the kitchen calendar and in the car kit. It should cover: contents check, stock rotation date, last practice drill, and the contact sheet for healthcare providers and neighbors. A laminated one-page medical plan for the household helps new caregivers step in without hesitation.

Finally, build a local support network that respects boundaries and safety. Share knowledge, not vulnerability. Exchange tips on what works, coordinate mutual aid for rides to clinics, and establish a quiet, neighborly channel for urgent needs when lines are busy or power is down. Resilience thrives where households lean on one another with competence and care.

EIGHT
HAND TOOLS, REPAIR SKILLS, AND KEEPING THINGS RUNNING

THE CORE HAND-TOOL KIT

A complete hand-tool kit is not a collection of fancy gadgets. It's a small, well-chosen set that sits on a bench or in a wooden box and can tackle wood, metal, fabric, and garden work without electricity. A sturdy toolkit saves time, reduces waste, and keeps a family moving when power lines are quiet. The practical rule is simple: fit your tools to the jobs you actually do, and choose quality you can trust for decades —not gadgets that wear out or become obsolete in a year.

Start with the basics for woodwork. A rip saw and a crosscut saw, both sharp and well-tuned, cover most sizing tasks. A good block plane and a smoothing plane make joints fit without struggle. A couple of chisels—from 1/4 inch to 1 inch—handle dados, mortises, and trimming. A marking knife, a square, and a marking gauge help you cut true lines. A quirk of old households is a fixed hand drill or brace and a box of center bits for precise holes in wood. Add a wooden mallet rather than steel for assembly, and a small rasp and file for shaping edges. Each of these tools is practical enough to stay in a single chest and durable enough to survive a storm stocking job after job.

For metalwork and hardware, the kit stays lean but capable. A hacksaw with a handful of blades, a few screwdrivers—flat and Phillips in common sizes—cover disassembly and assembly. A set of needle-nose pliers and locking pliers (vise grips) keeps stubborn joints from wandering apart. A punch and kicker set helps align fasteners. A wire brush cleans rust, and a small metal file handles burrs on cut edges. A hand drill with spare bits makes holes in metal or wood when a powered drill isn't available. This is where a modest investment pays off: tools that hold their edge, don't bend under pressure, and can be repaired if abused.

In the garden and around the yard, a practical collection includes a sharp billhook or pruning saw, a sturdy dibber, and a trowel that isn't flimsy. A hand cultivator loosens soil without fuel, while a cultivator fork breaks clumps without breaking your back. A good Dutch hoe or hoe-on-a-handle makes weed control efficient without scraping the soil too deeply. Hand shears for pruning and a sturdy watering can complement the set for maintenance tasks that keep a plot productive. The aim is not a wall of tools but a handful that you can trust to perform when you need them most.

Finally, a rugged, dependable storage system matters as much as the tools themselves. A wooden box, a metal tool chest, or a set of wall racks keeps everything accessible and dry. Labeling helps a household member locate the right tool quickly and reduces the chance of misplacing blades, bits, or gauges. A simple strap or band around bits keeps narrower tools from rattling loose. The organization you build today determines how fast a repair can begin tomorrow.

Concrete examples anchor this approach. A cabinetmaker in a nearby valley still uses the same brace and bit set his grandfather bought secondhand in the 1940s. He sharpens with a simple oil stone and a honing guide, then oils the blade edges every season. A small blacksmiths' bench in a rural workshop holds a hardy pair of tongs, a hardy anvil, and a straight-edge hammer. These are not museum pieces; they're everyday tools that keep a family productive for years. The toolkit that endures is the toolkit that's tended, slept with, and used regularly. Treat your set as a living system, not a shelf of memory.

The core kit is meant to be portable, repairable, and repeatable. It should empower you to start a job, finish it, and move on without waiting for a service call or a power outlet. When every household member learns to respect and rely on this basic set, a single day of work stops feeling like a bottleneck and starts feeling like progress.

SHARPENING, OILING, AND STORING TOOLS

Edges win battles. Dull tools lose them. The discipline is straightforward: keep metal sharp, keep joints tight, keep tools dry, and keep them easy to reach. A small routine can preserve an entire shop's usefulness through years of hard use.

Sharpening begins with a simple truth: a flat stone is forgiving, and a good bevel keeps a tool cutting where you want it. For chisels and plane blades, maintain a consistent bevel angle—roughly 25 degrees for most steel, with a slightly finer edge for finishing work. Use a two-stone system: a coarser stone (around 1000 grit) to establish or repair a burr, and a finer stone (around 4000–8000 grit) to refine and polish. A light film of honing oil on the stone helps particles ride away rather than clog the surface. For those without fancy stones, a well-worn oil-stone can do the job, as can a sharpening guide that holds the angle steady.

Knives, plane blades, and chisels respond to the same rule: even, deliberate strokes. Do not rush. Follow a consistent path, and test the edge with a soft shave or a light cut on scrap wood. When honing feels like guesswork, step back and realign the blade with a sight line or a square. The time you invest here pays off every time you pick up a tool and feel that decisive, clean bite.

Oiling is the second pillar. A thin film of mineral oil on blades and clamps prevents rust, slows corrosion, and makes sliding joints smoother. Wipe blades clean after use, then apply just enough oil to leave a faint sheen. Lightly oil moving parts—especially hinges on doors and lids—so they don't seize when humidity climbs. The goal is not slickness but survivability: tools must function when the world is damp or dusty.

Storage matters as much as technique. Keep tools in a dry place, off concrete floors where moisture rises. If you can, place a desiccant packet or a rag with silica gel in the box to absorb humidity. Keep sharp edges away from fragile finishes by using blade guards or wrapping blades with a clean cloth when stored. When tools sit for weeks, check for rust spots and wipe them away with a dry cloth and a quick dab of oil before you use them again.

A practical habit is the weekly wipe-down. It takes five minutes to inspect handles, check for loose screws, and oil metal surfaces. The payoff is measurable: tools that glide instead of grind, and a bench that stays productive rather than turning into a pile of excuses.

REPAIR BEFORE REPLACE

If you want to slow the pace of decline, you repair what you can before you replace what you can't. Small repairs are the common ground between a functioning home and a breaking point. The door hinges on a farmhouse cabinet, the tabletop joint that's loosened from years of use, or the thread in a cracked wooden handle—all are opportunities to practice doing the work rather than buying a new part.

Hinge repair is a first principle. A loose hinge can wobble a door and waste hours. Remove the hinge, clean the screw holes, and re-seat with wood dowels or longer screws on sturdy stock. If the screw holes are stripped, use a small wooden dowel coated in wood glue; drill out to a snug fit and remount. If you must, add a metal plate to reinforce the area. The goal is stability, not showy perfection. A door should swing true, not grind against the frame.

When joints loosen, glue and clamp become a household's best friend. Wood glue on a loose mortise, a dowelled repair to a chair leg, or a patch on a cracked rail can save an item that would otherwise be tossed. Dowels and a solid amount of glue minus excess give a durable, serviceable result. For furniture that bears weight, a pocket of epoxy resin can fill a worn gap and return load-bearing capacity without a full replacement.

Small hardware failures can be tackled with the same mindset. A stripped screw in a drawer pull is often a patch with a longer screw or a slightly larger diameter. If the screw holes are worn, fill with a bit of toothpick and glue, let dry, and reinsert. When a latch fails, replace the tang or re-thread the screw holes. Metal repairs—bent brackets, torn gaskets, or rusted brads—can be coaxed back into action with a little heat or the right punch. It's not about making things flawless; it's about keeping them functional.

The repair-first approach is not a shortcut. It's a discipline that slows costs, sustains households, and trains hands to diagnose problems instead of rushing to disposal. Each repair completed becomes a demonstration that a home is a working system, not a collection of fragile parts.

TEACHING PRACTICAL COMPETENCE AT HOME

Repair work stays alive when it's shared. Teaching practical competence at home turns maintenance from a one-time fix into a daily habit. The goal is ordinary, repeatable tasks that any capable household member can perform with a little guidance and time.

Create a simple sequence that becomes a family routine. Pick a weekend project that uses a handful of tools and yields a visible result: tighten loose drawer runners, replace a cracked handle, or re-hang a kitchen cabinet. Start with a short, clear set of steps, then let junior hands take the lead while an adult supervises. The act of guiding, not dictating, teaches more than the final outcome.

A durable repair curriculum uses seasonal cycles. In spring, focus on garden tools and deck maintenance. In late summer, repair or sharpen lawn equipment and refinish worn handles on frequently used gear. In fall, prep for winter with door seals, window putty, and chisel or plane work to fit new weatherstripping. A predictable cadence lowers anxiety and makes repair predictable, not extraordinary.

Document the process. A notebook with the date, the problem, the fix, and the tools used becomes a family atlas of competence. Children can sketch the steps on a simple diagram, then repeat it later. As skills accumulate, the same problems become easier to spot and fix. A home that teaches itself to repair grows into a home that refuses to surrender to breakdowns.

Finally, celebrate the practical wins. A chair that hasn't squeaked in months, a toolbox that opens smoothly, a cabinet door that stays shut—these are the small victories that build confidence. When a household uses its hands as a primary instrument, resilience becomes a steady, repeatable practice rather than a distant ideal.

NINE
CLOTHING, SEWING, AND DURABLE HOUSEHOLD GOODS

BUILDING A HARD-USE WARDROBE

Durable clothing is infrastructure you wear. When shelves thin out and online orders stall, the clothes you reach for become part of your defense. A well-made, repairable wardrobe does more than keep you warm or modestly dressed; it keeps a family moving. It enables work, care, and preparation without waiting for a shipment or a credit card to clear. Start with fabrics that tolerate hard use and ages well. Wool, duck canvas, denim, heavy twill, and oilcloth have the staying power of good tools. They resist wear, mend easily, and shed the kind of damage that ruins cheaper garments overnight.

In practice, that means choosing a small, repeatable set of pieces you can swap and patch rather than replace. A work shirt in dense cotton or flannel, a set of bib overalls or sturdy work pants, a wool or waxed-canvas jacket, an apron, sturdy socks, and a durable hat. Layering matters. A base layer keeps you warm when temperatures swing. A mid-layer insulates against wind. The outer shell shields rain and abrasion. This layering approach makes even modest materials serve twice as long.

Durability comes from construction. Look for double or triple stitching at stress points, reinforced patches at elbows and knees, and seams that lie flat enough to avoid snagging on rough surfaces. A straight-cut sleeve or leg reduces friction and pilling. Zippers last longer when backed by leather or canvas flaps, and snaps are preferable to fragile plastic hardware. In a bag or drawer, keep a small but steady stack of sturdy work shirts, a couple pairs of pants with reinforced knees, and a spare jacket that can withstand wind and damp.

Workwear should be designed for repair as much as for use. Keep color and fabric choices simple so patches and mends blend in. Mending becomes routine when you set aside a weekly window for it, not as a weekend luxury. Patch with purpose: a denim patch on a worn knee, a wool patch on a sleeve, or a leather elbow guard on a rugged jacket. Patch shapes should be deliberate and left long enough to weave in threads for secure anchoring. This is how a garment survives months of daily labor and a season of rough weather.

Closet organization matters. Keep a sturdy, moth-resistant storage plan: cedar blocks or lavender sachets, breathable cotton bags, and a clean space away from damp walls. Rotate seasonal garments so the ones used least get a rest period, reducing overall wear. Store shoes and boots in pairs, with laces tied and insoles in their natural state. If you can, designate a small washing-and-drying space that uses minimal electricity or none at all. Cleaning and airing garments after use helps fabrics last longer and stay smelling fresh, which matters when you're sharing space with others.

A few practical core pieces set the baseline. Two high-quality shirts in dense cotton or wool, two pairs of work pants or coveralls with reinforced seats, one heavy jacket or coat, one lighter jacket, one durable apron, and a couple of scarves or neck gaiters. Add one pair of sturdy boots or heavy shoes and a set of warm socks you can rotate. The goal is not a fashion statement but a reliable flow of clothing that can be repaired, mended, and extended to cover the longest stretch of disruption.

Keep care simple. Wash in cool water when possible, air-dry away from direct sun, and moisturize leather items as needed. A little oil on wooden combs, buttons, and buckles keeps them from drying out and breaking later. When you invest in the right fabrics and construction, you invest in fewer replacements and more consistent daily life. The payoff appears in cold mornings, late evenings, and the weeks when stores aren't an option.

The practical takeaway is clear: build a wardrobe that wears, not one that wears you down. Choose sturdy fabrics, favor repairable construction, and keep your most-used pieces available, organized, and ready to patch. A durable wardrobe reduces decisions, saves money, and buys time when every minute counts.

SEWING REPAIRS THAT SAVE MONEY

Repair is cheaper and faster than replacement when you build the habit. A torn elbow or a frayed knee doesn't have to become a lost garment. Hand sewing and small fixes extend life and preserve value, especially when new fabric or specialized gear becomes scarce. The core idea is to create an adaptable repair kit you can use without tools you don't own or a shop you can't access. You'll need a sturdy hand sewing kit: a sharp needle, thick thread in color families that match your garments, sharp scissors, a seam-ripper, a thimble, and a measuring tape. Add patches in materials you already own —denim scraps, leather scraps, wool or canvas—so you can patch without buying new fabric.

Begin with patches that cover high-wear areas. Elbows, knees, and seat seams take the brunt of daily work. A small patch is stronger if you cut it slightly larger than the worn area and rounded at the corners. Sew it from the inside for a cleaner look; the stitches should be dense and evenly spaced, about a quarter-inch apart. A simple running stitch with an extra loop around the edge helps the patch grip through repeated bending. For denim or canvas, use a heavy thread color-matched to blend in. For wool or leather, use a thread and needle suited to the fabric's density; a waxed thread can glide through thick fibers without fraying.

Hemming is another practical skill that saves money and time. A proper hem to restore length keeps pants workable for another season without changing the fabric at all. If you don't have a sewing machine, a strong, even blind stitch keeps hems secure and looks tidy from the outside. Patch a frayed edge rather than cutting it away if you can; this preserves fabric and reduces the risk of further unraveling. Replace broken buttons with sturdy shanks; when buttons are scarce, use rivets or small patches of fabric as a temporary anchor, then sew on a proper button when it's available again.

Socks, stockings, and soft goods deserve attention, too. A simple darn can extend the life of a sock that has developed a run or hole. Darning is easiest when you align the fabric and weave new threads across the hole in parallel lines to recreate the fabric's original structure. For knit items, use a zigzag stitch if you have a machine; by hand, weave the threads in and out so the fabric remains flexible rather than stiff. Replace elastic bands in waistbands, cuffs, and sleeves when they sag; modest, well-made elastic can stretch a lot of life back into garments that would otherwise be thrown away.

Your repair repertoire should include quick fixes you can perform with minimal tools. Add a few improvised options for differing fabrics: a leather elbow pad for a rough-work jacket, a cotton patch for a seam that keeps pulling, or a small piece of waterproof tape on a rain shell until you can stitch or patch properly. The point isn't to look perfect; it's to keep working gear on your back and on your feet. When you treat repairs as a normal part of upkeep, you save money and reduce dependence on outside markets.

A practical repair habit matters as much as the patch itself. Set aside a dedicated hour each week for stewardship of garments: mend the worn spots, resew the hems, replace a button, refasten a strap. Keep record of what you repaired and what you replaced. The list becomes a guide for future purchases, helping you choose fabrics that resist wear in the exact places your family uses most. Repair culture is discipline in motion; it keeps a household moving forward even when the supply chain falters.

BEDDING, TOWELS, AND HOUSEHOLD TEXTILES

Durable textiles are the quiet backbone of a home in lean times. Linen, wool, cotton, and cotton blends endure with less fuss than delicate fabrics. The idea is simple: plan for the long haul by stocking items you won't have to replace every season. Start with core sets: two or three sturdy bed sheets per person, a couple of blankets, a handful of heavy towels, and a set of table linens that can handle daily use and a few mishaps. Look for dense weaves that resist pilling, and fabrics that improve with age rather than degrade. Wool blankets and solid cotton sheets survive frequent washing and rough handling better than shawls or lightweight fabrics.

Storage matters as much as selection. Keep textiles in a dry, dark space where pests won't snack on them. Use breathable cotton or linen storage bags rather than plastic, which can trap moisture and invite mildew. Place cedar blocks or lavender sachets in bins to deter moths and beetles. Rotate textiles seasonally: summer sheets, winter blankets, and spare towels move to top shelves so you don't forget what you have. Clear labeling helps a family of different ages find what they need without rifling through boxes.

Maintenance is part of the plan. Wash linens in gentle cycles or by hand when possible. Air-dry to preserve fibers; heat can accelerate wear and shrinkage. For towels that feel stiff after years of use, a quick wash with a little white vinegar during the rinse cycle can restore softness and remove mineral buildup. Stains should be treated at once with a simple paste of baking soda and water or with a mild soap. If you must bleach, do so sparingly and with proper dilution, because harsh chemicals degrade fibers over time and shorten lifespans.

Textile longevity depends on smart use. Don't keep a surplus of low-quality towels or sheets that wear out fast. Prioritize weight, weave, and fiber quality over flashy colors. A few durable textiles, maintained through routine washing and gentle drying, will outlast a dozen cheaper items. The goal is to keep enough sturdy textiles on hand so a family can wash and rotate without an urgent trip to the store.

In addition to bed and bath, essential textiles extend beyond dress and linen. Kitchen towels, cloth napkins, and cloth diapers or wipes become routines that reduce waste and dependence on disposable goods. A simple, washable napkin reduces the need for single-use paper products and keeps meals orderly during a disruption. By choosing robust textiles, you create a reservoir of coverage—protecting clothing, bedding, and daily needs during weeks or months without a steady supply line.

A practical rule holds here: durability is a habit of care as much as a property of fabric. Store textiles dry, clean, and properly spaced. Inspect them after each wash, mend as needed, and consider the lifecycle of each fabric—how many years of service it has left and how easily it can be repaired or repurposed. The more you plan for the long haul, the less you'll fear the moments when the grid goes quiet.

LEATHER, CANVAS, AND BOOT CARE

A single good boot saves you the cost of several cheap pairs. Leather, canvas, and other durable materials demand attention if they're going to serve months of hard use. Start with the basics: clear dirt off leather or canvas with a soft brush, then wipe down with a damp cloth. Let things dry completely before applying conditioner or waterproofing. Leather loves ordinary care—clean, moisturize, and protect against moisture. A simple conditioning routine, using a natural oil or a dedicated leather conditioner, keeps fibers supple and reduces cracking. If you live in damp climates, apply a thin layer of wax or a wax-based conditioner to seal out moisture. This is not a one-off fix but a recurring habit that pays dividends over a long cold season.

Waterproofing is a key advantage of durable goods. A wax-based waterproofing treatment on boots, jackets, and canvas sacks can turn ordinary gear into weatherproof work horses. Reapply after heavy use or washing, and store treated items away from heat that could dry out fibers. Waxing can be done with a simple scrub and melt approach or with a dedicated product that adds a lasting, breathable layer. The goal is not to create a impermeable shell but to slow down water infiltration while keeping the fabric flexible.

Repair keeps high-value items in circulation longer. A boot sole that begins to separate can be reinforced with stitching or a patch of leather or rubber. Replacing worn out laces is a simple, cheap improvement that makes an old pair usable again. Canvas tents, tarps, and heavy deerskin gloves benefit from extra stitching or a patch of strong fabric when seams start to fail. With leather goods, check for cracks and rotate the item that receives the most wear to avoid early failure of a single product.

Organization matters here as well. Store boots and heavy gear in a dry, wide space that allows air to circulate. Keep nails and rivets where you can access them quickly, and have a small toolbox ready for quick fixes. When you invest in one reliable, well-made canvas bag or sheath, you gain a versatile container for feed, tools, or groceries—another example of making durable goods multiply in usefulness. A well-cared-for leather item becomes a family heirloom in time, proof that durability can be practical and personal, not decorative.

The takeaway is practical and clear: invest in sturdy fabrics, repair them faithfully, and treat your gear with a routine of cleaning, conditioning, and protection. Durable textiles and leather don't just survive. They enable a household to stay active, fed, and warm when external systems fail. By treating clothing, fabrics, and leather as essential infrastructure, you build a self-reliant rhythm that holds steady through long days of disruption and quiet nights alike.

TEN
LIVESTOCK, BACKYARD ANIMALS, AND USEFUL PROTEIN

PRESERVATION, ROOT CELLARING, AND EATING THROUGH WINTER

Preservation is the last line of defense against hunger. A harvest without a plan spoils, but a harvest with a plan outlasts the cold months. When the grid is quiet for weeks at a time, the difference between a thriving pantry and a bare table comes down to storage: how you keep food from spoiling, how you stretch it, and how you choose to use it. This chapter treats preservation not as an afterthought, but as a core skill—a practical discipline that protects health, reduces waste, and keeps meals steady when markets fail and electricity vanishes. The method you rely on most should fit your land, your skills, and your patience. You don't need fancy equipment to create reliable winter food, just reliable routines and sturdy containers that you trust year after year.

Storage is a system, not a miracle. You build it with the same care you'd give a root crop or a clean tool handle: a defined space, a way to move air, and a schedule for checking and rotating. The goal is simple: keep taste, texture, and nutrition intact as long as possible without relying on a refrigerator or a charger. Think of it as a small factory inside the home where everyday actions—washing, packing, labeling, moving food to where it will stay cool—become the backbone of daily life during shortages. Good storage begins before the harvest. It requires planning, not luck.

Root crops, apples, greens, and even meat can survive long winters if they are kept in the right conditions and handled with care. The best storage conditions aren't extreme. They are modest, steady, and experienced. A root cellar or a cool, dark corner of the house can carry you through months of lean days. The emphasis here is versatility. You'll combine several methods to cover different foods and different rhythms of usage. A winter pantry is a living inventory—one that demands attention, but pays back in reliability and peace of mind. The first rule is to store what you actually use and use what you store. Waste is the enemy; consistency is the ally.

What follows are practical, field-tested methods that work with minimal electricity and maximum simplicity. You'll learn to choose the right storage for your space, to prepare foods for long life, and to create a winter routine that keeps your kitchen productive without constant trips to the store. The emphasis is on food you already know how to grow, harvest, or catch, and on keeping it safe and tasty through the cold season. Your goal is not perfection, but dependability.

Framing storage around your land and your life

The heart of preservation is choosing where and how to store. A root cellar that stays near the mid-40s Fahrenheit with high humidity is a dream, but any reliable cool space can do. If you have a basement, it can serve as a cellar with a few adjustments—ventilation, insulation, and careful organization. If you don't, a shaded crawl space, a cool outbuilding, or even a north-facing interior wall with a thick display of crates can work. The key is consistent temperature and steady humidity. Food stored in a spot that swings between warm and cold ages faster and spoils sooner.

Your first step is to map what you'll store where. Potatoes in one corner, carrots and beets in another, apples or pears in a deeper, darker shelf. Green vegetables, cabbage, and onions demand a different humidity level than dried herbs or fruit preserves. Draw a simple layout and label shelves for what goes where. Rotation is your friend: move last month's stock to the back and bring forward what's older. The discipline of rotation is the difference between a winter where you comfort-eat your own preserves and a winter where you eat what you need without waste.

Nothing spoils a plan like poor sanitation or contaminated containers. Clean jars, lids, and tools stay on hand year after year. DOn't reuse questionable jars or lids from uncertain sources. Keep your packing clean, dry, and clearly labeled. A simple labeling system—date, contents, storage method—reduces waste and confusion when the pantry grows crowded.

Finally, build in checks. A quick once-a-week walk-through to check for soft spots, odd odors, or moisture helps catch problems before they spread. A small amount of spoiled product can ruin a larger batch if not caught early. The steady hand of routine keeps your storage healthy and reliable.

The end result is a home that feels predictable even when the weather isn't. You know where your food lives, how long it should last, and what to do when it doesn't. That predictability reduces anxiety and makes the long winter less daunting. Preservation becomes a habit, not a plan you scramble to implement when the first frost hits. You win by keeping your shelves full and your meals honest.

A spectrum of storage methods you can rely on

Electricity isn't the only way to keep food safe. The methods you'll use are varied and complementary, allowing you to tailor storage to what you have and what you can harvest.

Root cellaring remains the backbone. A cool space with high humidity can cradle root crops and winter greens for months. Carrots lose their snap but stay sweet when kept damp; beets stay firm; parsnips mellow and sweeten with time. Potatoes are perhaps the most forgiving, storing well in a dim, cool location with some air movement. Cabbage can survive well in cooler, more humid air, and apples, pears, and other fruits can ride through a long snap of cold days if kept dark and cool. The mix of vegetables and fruit in a root cellar is where you'll see the most durable yields.

Canning lets you lock in flavors and textures by turning fresh vegetables, meats, and sauces into shelf-stable jars. Water-bath canning works for high-acid foods—salsa, pickles, jams, and fruit products—while pressure canning is necessary for low-acid items like vegetables and meats. A compact, inexpensive canner, plus a handful of jars and proper lids, becomes a small time capsule that holds summer well into winter. The rule is simple: use tested recipes and follow safe processing times. Don't improvise unsafe methods. Save your notes and date every jar so you rotate through them.

Drying removes moisture, the single biggest stickler for spoilage. You can sun-dry herbs, peppers, and fruit, or use shade-drying for foods with higher moisture. An oven can serve as a controlled drying chamber if you use low heat and a vented door. A dedicated dehydrator is handy if you have a lot of greens or fruit, but it's not a necessity. Dried foods take up less space and are easy to portion. When you rehydrate, you know exactly what you're getting—textured foods not far from their fresh selves.

Fermentation preserves and adds depth of flavor. Sauerkraut, pickled cabbage, kimchi, and brined vegetables stay edible for months in a cool corner. Fermentation doesn't require electricity or fancy equipment. It does require salt, clean water, and a container that can hold up under pressure as fermentation gas builds. The result is not only safe storage but a complex, satisfying taste that gets people talking about what's in the pantry.

Finally, consider salt-curing and smoke for meat, fish, or larger cuts when you have the means for a basic smokehouse or a simple salt cure in a cool space. Salt draws out moisture and creates an inhospitable environment for spoilage organisms. If you go this route, ensure you have clean meat, proper salting, and a way to prevent mold and off-flavors. These methods extend your options beyond fresh harvests and give you flexibility when supplies are tight.

The bottom line is simple: pick two or three storage methods that suit your space and your harvest, then build a routine around them. You'll create a resilient winter pantry that doesn't rely on an outage-proof grid and keeps meals grounded in what you know how to grow and raise.

Root cellaring in practice

A root cellar can be a hillside space, a basement corner, or a purpose-built, lightly buried chamber. The goal isn't a perfect replica of a commercial cold store; it's a steady, low-energy environment with reliable airflow and humidity. Start with the basics: shade, cool, and moist air. Ventilation is essential. A small vent at the top and another near the bottom creates a gentle draft, which keeps air moving without chilling a space that's already cold. If you're using an improvised space, block radiant heat sources—pipes, furnaces, or hot water lines—to prevent temperature spikes.

Shelving should be adjustable and strong. Crates with slatted bottoms work well, allowing air to circulate around roots. Vegetables go in crates labeled by type. Carrots, turnips, and beets ride in damp sand or damp straw to maintain humidity. Potatoes stay away from onions and apples; the ethylene gas from certain fruits can hasten spoiling and sprouting in tubers.

Apples keep best in a different corner where they'll stay cool and dark, with a separate humidity balance. Each crop has its own rules, and keeping them separate prevents cross-impact.

Check on your cellar weekly. Remove any signs of spoilage; wipe moisture from surfaces; rearrange stacks so air moves freely. A handful of lids or old cheesecloth can cover openings to keep pests out while still letting air move. A modest humidity, in the range of roughly 85–95 percent, protects greens and root vegetables from shriveling while not encouraging mold. Temperatures around the upper 30s to mid-40s Fahrenheit keep these crops fresh for months.

Your cellar won't need constant human attention, but it will reward regular care. When you know where to go for a given crop, you reduce waste, speed meal prep, and maintain a steady rhythm through winter. A good root cellar becomes a quiet, reliable partner in daily life, delivering predictable sustenance when modern systems falter.

Safe, sure canning, drying, and fermenting routines

Canning isn't a one-off project; it's a routine that pays out with steadier meals. Start with familiar foods you already enjoy: tomato sauce, beans, peaches, pickles. Confirm you're using safe, tested recipes. Label every jar with contents and date so you know what you're pulling from the shelf in a pinch. If you're new to pressure canning, practice with a small batch and follow the manufacturer's instructions to the letter. The goal is simple: prevent botulism and ensure consistent quality across the year.

Drying requires clean heat and steady air. Check foods for doneness and condition before storing. The moment you spot moisture or mold, you discard. Dry thoroughly and store in airtight containers away from sunlight. Recheck your containers occasionally; moisture can creep in through imperfect seals.

Fermentation turns living cultures into preserved flavor. It's forgiving but not lazy. Salt helps preserve texture and flavor; use clean vessels, keep the product submerged, and monitor for off smells or unusual colors. A successful batch yields a tangy, crisp result that's both nutritious and satisfying.

Bundle your preservation methods with a personal inventory system. Keep a simple ledger of what you have, what you used, and when you expect to dip into a given stock. The more you can rely on your own products, the less you'll worry about shortages elsewhere. And if you ever doubt your ability to eat through winter, remember this: well-managed food storage multiplies your options, not your worries.

Planning meals around preserved foods

Winter meals are built on what's on hand, not what's freshest in the market. A box of potatoes, a jar of tomato sauce, a bag of dried beans, a bottle of kraut or pickle, and a stock of dried herbs can anchor a week's worth of dinners. Plan around a weekly cycle: two days with hot soups or stews built from root vegetables; one meat option that uses preserved products; a couple of meals that rely on stored fruit and dried grains. The discipline of planning reduces waste and improves nutrition, even when fresh groceries feel scarce.

Craft a monthly menu that rotates the same core ingredients. If you have apples, plan a couple of desserts or savory dishes that feature them. If you have cabbage, plan a kraut success story, a stuffed dish, and a slaw that can accompany many meals. The aim is to minimize the number of times you open cold storage and maximize the use of your preserves. A well-balanced rotation guarantees that you don't lose track of what you have and what you need.

Finally, share the surplus. If you have more jams, pickles, or dried goods than your family can consume, trade with neighbors or stock a small barter shelf. A predictable, modest exchange can sustain relationships, keep prices stable, and reduce waste. Winter living thrives on social connections as much as on jars and root bins. A robust plan turns a season of restriction into a season of abundance, and that's the core of resilience.

ELEVEN
PRESERVATION, ROOT CELLARING, AND EATING THROUGH WINTER

PRESERVATION, ROOT CELLARING, AND EATING THROUGH WINTER

Preservation isn't a luxury; it's the foundation of a home that can weather weeks of emptiness. When the grid falters, what you can't grow, you must have saved already. The trick isn't a fancy gadget. It's reliable methods, practiced routines, and a pantry that behaves like a backyard harvest you can reach year-round. This chapter shows how to keep food safe, tasty, and usable from one harvest to the next using water-bath and pressure canning, drying, fermenting, smoking, root storage, and disciplined inventory.

WATER-BATH AND PRESSURE CANNING

Canning is the quiet backbone of a food plan that doesn't rely on electricity to stay alive. Water-bath canning works for high-acid foods and close-topped jars quickly enough to preserve fruit, jam, pickles, and tomato products. Pressure canning handles low-acid foods—meats, poultry, fish, vegetables with pH above 4.6—by turning water into a higher-pressure boil that safely reaches the temperatures needed to kill botulism-causing organisms. For households aiming for long, steady resilience, both methods have a place, and both demand respect for safety, precision, and cleanliness.

Start with the basics: clean jars and two-piece lids, kept hot until use. Jars should be free of nicks and cracks; lids must seal properly—no wobbly lids, no rusty rings. You'll need a canner with a rack for water-bath work or a dedicated pressure canner. A simple checklist helps: washed jars, lids in hot water but not boiling, clean bands, a ladle, a jar lifter, a timing device, and a sturdy counter space. Labeling is not optional; a dated label on each jar keeps you honest with memory and rotation.

High-acid foods require only a water bath. Think fruit compotes, peaches in syrup, applesauce, jams, jellies, and tomato products acidified to ensure safe pH. The acid isn't just flavor; it's a safety buffer. Add lemon juice or a pinch of citric acid to tomato products if you're canning without a tested recipe. Process times hinge on jar size, food density, and altitude. At sea level, a standard pint jar of peaches may process for 20 minutes; for tomato sauce, 40 minutes. At 2,000 feet, add a few extra minutes; at higher elevations, follow a tested altitude-adjusted chart. Always ensure there's enough water to cover jars by at least an inch and that the water returns to a rolling boil before the timer starts.

Low-acid foods demand pressure canning because ordinary boiling can't reach temperatures high enough to destroy dangerous organisms. Ground meat, poultry, seafood, beans, corn, carrots, and mixed vegetables fall into this category. Your pressure canner acts as a controlled pressure vessel. Build a basic routine: heat the canner gradually, exhaust air as directed by your model, seal it, bring to the required pressure, and time precisely. When you're done, cool gradually; venting too quickly can cause seals to fail or jars to crack. Store sealed jars in a cool, dark place and rotate with your other preserved goods.

A few practical anchors make canning safer and more predictable. Use tested recipes from reliable sources—extension services, university co-ops, or experienced canners in your network. Never improvise low-acid canning with untested methods. Maintain clean work surfaces and avoid cross-contamination with raw and cooked foods. If a seal fails, refrigerate the contents if you'll use them within a few days; otherwise, discard to prevent foodborne illness. Durable jars and reliable lids are less expensive than the cost of waste.

Water-bath and pressure canning together give you a shelf-stable reservoir that doesn't tie you to energy as a necessity. The goal isn't perfection but predictability: a pantry that keeps taste and nutrition intact through months when the market is unreliable. A well-run canning program also teaches you timing, sanitation, and methodical work—habits that transfer to other preservation tasks.

DRYING, FERMENTING, AND SMOKING

Drying is the simplest, oldest way to store harvests. It's light, portable, and forgiving. Herbs dry in a sunny windowsill;

peppers, tomatoes, apples, and pears can dry in a warm room or a dehydrator. The test for doneness is moisture: a piece should bend without sweating, not feel rubbery or soft. When fully dried, foods should be stored in airtight containers in a cool, dark spot. Keep a label on each batch with date and content. You'll be surprised how a jar of dried mushrooms or oregano can stretch a winter's cooking, much as a handful of dried fruit can sweeten a morning meal.

Fermenting turns a harvest into probiotic-rich staples that stay useful long after fresh produce fades. A head of cabbage becomes sauerkraut with salt and time; cucumbers mellow into pickles when kept submerged under brine. The magic is simple: salt draws out moisture and creates an environment where beneficial microbes can thrive. A clean jar, a little salt, a lid that lets gas escape or a fermentation weight, and a cool corner are all you need. Fermentation rewards patience. Weeks later, you'll taste depth—more complex than any quick pickle from the store. Fermented foods stay eatable for months if kept cool and hidden from direct light.

Smoking food concentrates flavors and lengthens shelf life. Hot smoking both cooks and preserves; the smoke coats meat or fish with compounds that deter spoilage. Cold smoking needs more caution, especially in nonindustrial setups; it adds flavor without cooking. If you build a small smokehouse, choose hardwoods suited for the meat and fish you'll preserve. Manage temperature carefully, and finish with a quick cook to safe internal temperatures when needed. Smoking is a craft that rewards consistent wood use, temperature control, and humidity awareness. A well-smoked sausage or fish can carry dinner through weeks with fewer groceries and less waste.

Drying, fermenting, and smoking are not nostalgia; they're flexible, scalable tools. They let you lean on taste, texture, and nutrient density while avoiding reliance on refrigeration. Each method complements canning by providing a range of textures and flavors that keep winter menus interesting, affordable, and safe.

ROOT CELLARS AND COOL STORAGE

A well-tuned root cellar starts with the right climate. You want darkness, stable cool temperatures, and humidity that keeps produce from drying out or molding. If you're lucky enough to have a basement or a cool interior corner, you can carve out this space. If not, a shaded, unheated closet or a sheltered external storage area can approximate the room. The target is a space hovering near 32–40°F (0–4°C) with humidity around 85–95% for most roots and greens. A thermometer and a hygrometer become your best friends.

Potatoes, onions, carrots, beets, cabbage, apples, squash, and applesauce jars are prime candidates for root storage. Potatoes go in burlap sacks or wire baskets, layered with straw to maintain airflow. Onions hang in mesh bags or braided strands where air can move freely. Carrots and beets keep best in bins of damp sand or sawdust; the moisture slows dehydration without promoting rot. Apples tolerate cool, humid air but need regular checking for bruises and early rot. Squash, pumpkins, and gourds store well in crates with space between each piece and no moisture pockets where rot can hide.

The trick is keeping produce away from ethylene producers and storing items that mature at different rates apart. Apples and potatoes can influence other vegetables, accelerating spoilage if kept together in poor air flow. A small fan can help in larger spaces, but even a simple, well-ventilated shelf system works if you rotate weekly. Check every week for soft spots, mold, or sprouting; remove spoiled items immediately to protect the rest. A root cellar is not a museum; it's a working backbone that keeps staples stable through winter.

INVENTORY CONTROL AND ROTATION

If storage is a habit, waste becomes a rare anomaly. Inventory control starts with a clear map of what you have, where it lives, and when you should reach for it. First In, First Out. You want the oldest items at the front, the newest tucked behind, so nothing lingers past its prime. Label every jar, bag, and bin with the content and date. A simple ledger or spreadsheet can track quantities, but a physical label method—chalk on shelves, or colored tape—works just as well in a pinch.

Create a storage map. List zones for canned goods, dried foods, root storage, and special items like maple syrups or honey. In each zone, place a small card or tag that reminds you of typical shelf life and best uses. Monthly checks keep accuracy honest. When you harvest in 夏, you'll have new items; when you dine, you'll know exactly what's available without opening every jar.

A culture of rotation reduces waste and builds confidence. If the family knows where to find oats for breakfast or a jar of applesauce for a sick day, the household runs smoother. This isn't about being perfect; it's about being prepared enough to feed the family with minimal trips to the store. A well-managed pantry also helps you stretch dollars by preventing spoilage and encouraging the use of older staples in daily meals.

The end result is a living system: a pantry that supports meals, a cellar that stabilizes winter, and a set of routines that make feeding a household through tough months feel ordinary rather than extraordinary. You'll find resilience isn't a dramatic event. It's a practiced, steady discipline that keeps shelves full and minds calm.

TWELVE
SEEDS, SOIL, AND THE NEXT SEASON

Soil is more than dirt. It is a living engine for resilience, a quiet partner in every harvest, and the only true hedge against rising food costs. A garden that treats soil like a partner instead of a backdrop can outproduce most store shelves, even when power and supply lines fail. The habit to cultivate is simple: feed the soil, save what grows well, stretch those greens with clever protection, and plan backward from harvest toward next year. Do this, and a season becomes a loop of abundance rather than a race for the next shipment.

SOIL FERTILITY ON A BUDGET

Healthy soil isn't expensive, it's intentional. The backbone of a resilient garden is a steady rhythm of feeding, turning, and protecting the ground you plant in. Start with a basic rotation plan. Beans, corn, and squash each pull a different nutrient load from the soil; rotating them keeps soil chemistry from tipping toward exhaustion. If you only have a few beds, rotate by bed color or by row; keep robust nitrogen users separate from nitrogen-hungry crops and reserve a bed for a legume that can fix nitrogen for the next round.

Compost is the cheapest, most versatile fertilizer you'll build. A simple hot or cold pile will do, but the more you mix and the more you turn, the faster you'll see results. Kitchen scraps, fallen leaves, grass clippings, and spent plants from last year all become soil gold when they break down. A pile that's too wet or too dry won't break down well; aim for a damp, crumbly texture that feels like a wrung-out sponge. In time, this becomes dark, sweet-smelling matter you can work into beds or mulch on top of seedlings.

Manure is another powerful ally, provided you source it responsibly. Fresh manure can burn tender roots, so let it age or compost it with straw or leaves to temper its heat. Poultry litter is especially potent, but it also carries more salt and weed seeds. Mixing a wheelbarrow of aged manure into a bed before sowing seeds gives you a quick mineral boost and improves soil structure. If you don't have access to manure, seaweed or kelp meal, bone meal, and rock phosphate can supplement phosphorus and trace minerals in smaller doses.

Mulch acts as a shield for soil life and a starter for moisture retention. Straw, shredded leaves, or hay laid on a 2-4 inch layer around plants reduces moisture loss, keeps weed pressure down, and moderates soil temperature. Mulch is cheap, easy to replace, and can be produced on site from fallen leaves and yard waste. The effect shows up as steadier germination, less soil crusting after rain, and healthier roots.

Mulching also ties into a practical pest strategy. A generous mulch blanket makes it harder for cutworms to reach emerging seedlings and helps keep slugs from starving them out at the stem. It's not a guaranteed fix, but it reduces pressure and buys time for natural predators, like toads and birds, to work.

A simple, repeatable routine makes soil health a habit rather than an experiment. Each beds cycle can be scheduled: spread compost in early spring, lay down mulch after transplant, and feed with a liquid compost tea mid-season. A dedicated space for composting and a selection of cover crops—vetch, clover, rye, or field peas—keep soil biology active through the off-season. Cover crops break soil crusts, scavenge residual nutrients, and, when tilled in, contribute organic matter that improves soil structure.

Healthy soil also means water efficiency. Compost improves water-holding capacity, so drought weeks won't collapse plant vigor as quickly. In contrast, poor soil drains away moisture, leaving roots stressed and pests ready to move in. The rule is simple: nourish the soil, and it nourishes your plants back. The payoff is a more forgiving garden that can withstand a late frost or a dry spell without collapsing. The ultimate budget winner is soil that works without constant inputs.

The end goal is a garden that produces with less fuel, fewer external inputs, and more natural cycles working in your favor. When you build soil steadily year after year, you're not just growing food; you're creating an evolving, resilient platform for your household to lean on when the grid shakes. Healthy soil becomes the quiet backbone of a family's year—low maintenance, high return, and surprisingly forgiving.

SEED SAVING FOR BEGINNERS

Saving seed is a practical act of faith in your own land. It is how a garden becomes a self-reinforcing system: you plant, you save, you plant again, and the line of better-adapted varieties grows stronger with every season. Start small with open-pollinated crops that reliably retain their traits from year to year. Hybrids may promise vigor, but saved seeds from hybrids won't produce the same plant or fruit, and that can be frustrating when you're trying to close gaps in your calendar.

Beans and peas are among the easiest for beginners. Dry beans, when left to mature on the vine, can be harvested once the pods turn brown and rattle. Dry them further before storage, then shell and keep in labeled jars. Peas are similar, with the added advantage that you can shell and dry the seeds when the pods dry to the point of snapping. Sunflowers, if you let the heads dry on the stalk, reveal seeds that are forgiving to save and plant. They also draw pollinators and create a bird-friendly surplus.

Tomatoes, peppers, and squash offer open-pollinated options that store well and reproduce true to type with careful handling. For tomato seeds, separate fruit with mature, pulpy seeds, ferment for a few days, then rinse and dry. The fermentation step kills off weak seed and helps prevent mold. For peppers, wait until fruits are fully mature and dry, then harvest and dry the seeds. Squash and cucumber seeds simply dry on a screen or rack away from direct sun. Label everything with plant name, year saved, and any notes about pollination proximity to other varieties.

The big idea is to keep things simple. Choose crops you already grow reliably, have a known taste and texture your family enjoys, and that store well. A small seed-saving plan can be written on the back of a notebook page: crop choices, isolation notes, harvest windows, drying locations, and storage containers. Glass jars with airtight seals in a cool, dry place are ideal. Label with year and variety and rotate stocks so your oldest seeds are used first. Germination tests—saving a handful of seeds for a week or two and then trying to grow them—are cheap and revealing.

Seed saving is also a family project. Children learn patience watching seeds change into plants, and adults sharpen records-keeping and observation skills. In a long-term disruption, a single saved bag of beans multiplies into meals, fodder, and future crops. You're building an archive that feeds the next season and teaches the next generation how to adapt with the land.

EXTENDING THE GROWING SEASON

A cold season in the garden is not a failure; it's a challenge to outthink the calendar. The simplest ways to extend your growing season are low-tech and repeatable: row covers, cold frames, and careful timing. Row covers—low hoops topped with translucent fabric—create a microclimate that protects seedlings from frost and pest damage while still letting in sunlight. They are inexpensive, removable, and durable enough to last several seasons.

Cold frames turn a nook of the garden into a miniature greenhouse. A frame built from salvaged lumber or bricks with a reused window on top can push the last frost boundary a week or two in spring and extend harvest into late fall. The best cold frames are sturdy, well-sealed, and easy to ventilate when sun warms the day. If you lack a clear window, a simple plastic-covered structure using PVC hoops will still work, particularly for greens, spinach, and radishes that can tolerate cooler soils.

Timing matters more than gadgets. Plant frost-tolerant crops early, then stagger planting by two- to four-week intervals to spread risk and keep the harvest coming. A tidy soil bed with a mulch layer warms faster and dries out slower, giving you a comfortable window to push roots deeper without sacrificing germination. When the forecast hints at a hard freeze, you can pull the covers over top and keep roots alive under a thin blanket of warmth from the soil itself.

In practical terms, a single season can be extended by a few extra weeks of leafy greens or by squeezing in a late-summer harvest of root crops after a fall rain. The trick is not grand machinery but a few deliberate steps: prepare the frame, choose the right crops for late-season production, and adjust spacing so plants don't compete in crowded quarters. A well-timed cover and a ready frame can turn a marginal month into a productive stretch, enough to make a real dent in the pantry deficit.

PLANNING NEXT YEAR WHILE HARVESTING THIS YEAR

Harvest time is planning time. Each bed you harvest becomes a notebook entry, a data point, and a seed map for the coming season. Record what grew well, what failed, and what surprises showed up in the soil. It's tempting to chase brightest blooms or biggest fruit, but resilience favors predictability: you want crops that survive your soil, your climate, and your water cycle with minimal fuss.

A simple rotation plan can be sketched on graph paper or a chalkboard calendar. Note plant families and their typical nutrient needs, then assign beds so no bed hosts the same family two years in a row. A four-bed rotation works for many households: legumes to fix nitrogen, root crops that break soil, leafy greens, and fruiting vegetables. The goal is a steady, repeatable rhythm rather than a grand overhaul each year.

Seed saving breeds continuity. Identify the crops your family loves most and start saving seeds from those varieties. Keep a seed bank log: variety, year, storage method, germination results, and notes about isolation if you're growing multiple varieties in proximity. A tiny investment in durable jars and silica packs can stretch your stock for years, preserving flavor and texture when external systems falter.

Planning also means logistics and roles. Who plants, who harvests, who saves seed, and who keeps the records? A simple schedule with a few household tasks spread across the week creates a rhythm. In a long disruption, routine is a form of security: predictable chores, reliable inputs, and people who know how to fill gaps when someone is away. The plan should accommodate weather surprises, soil fatigue, and family schedules, not pretend perfect conditions exist.

Finally, the bigger picture matters. The seeds you save and the soil you steward today become the backbone of your next season's garden and the community you rely on for barter, advice, and shared labor. A garden is not just food; it is a living system that repeats and improves with care. The result is a cycle that strengthens your household and deepens your roots in the land you tend.

THIRTEEN
HOUSEHOLD ECONOMY, BARTER, AND LOCAL TRADE

Cash may lose value, but useful goods and trusted skills keep moving. A family that learns to live with less cash does not become poorer. It becomes more predictable. In hard times, those with reliable routines, steady output, and a hand that reaches out to neighbors don't wait for a store to reopen. They trade what they can make for what they need, and they do it without drama. This is the core of Household Economy, Barter, and Local Trade: build value locally, earn trust, and let that trust carry you through shortages. The better you know your neighbors and your own crafts, the less you depend on distant supply chains. You become the kind of household others want to rely on in a crisis, and that reliability is money in good times and bad.

RUNNING A LOW-CASH HOUSEHOLD

Thrift isn't a ritual; it's a workflow. It starts with a simple, stubborn idea: spend on essentials, not on novelty. That means planning meals around what you already have, what's on sale, and what you can produce or preserve. A household that eats well on little cash runs three core habits: deliberate planning, waste reduction, and proactive maintenance.

Planning begins with a weekly menu built from staple ingredients—rice or beans, oats, dried or canned vegetables, fats, and a protein that fits your space and climate. Then you map what you already possess: jars of jam, dried beans, three-quarters of a bag of flour, leverage on-hand spices. Next comes rotation: older staples are used first, fresh purchases wrap around them, and nothing sits unused for too long. This keeps the pantry lean but effective. It also clarifies when you truly need to buy more and when you don't.

Waste is where cash evaporates the fastest. Cook in batches. Reheat leftovers rather than start anew. Use veggie scraps for broth, rinds for flavoring, cores for stock. When produce fails to get eaten, you don't toss it—conserve it. Freeze or dry where possible; even a small dehydrator can turn peppers, apples, and herbs into long-lived fuel for flavor. A family can stretch a month's groceries by 20–40 percent with careful planning and disciplined reuse of every edible fragment.

Maintenance is the unsung thrift discipline. A leaky faucet, a cracked window seal, a clogged gutter—those are the little expenses that compound quickly. A weekly, 30-minute check of hoses, outlets, seals, and filters protects against larger, unplanned costs. The goal isn't perfection; it's predictability. A predictable budget buys peace of mind when prices swing or supply lines falter. A practical approach is the cash envelope: allocate a fixed amount for groceries, fuel, and household basics, and use only that money for those categories. If you don't spend it all, you roll it into the next period; if you do, you adjust the following week. This is how a household guards against the habit of overspending in uncertain markets.

Another decisive habit is to produce more at home where possible. A small garden, even in a city yard, can give you greens, herbs, and a steady trickle of tomatoes or peppers. A few lays of crops can save money all year round, and root crops stored in a cool, dark place can be valuable barter items when cash is tight. Canning, drying, or fermenting preserves the harvest and creates tradable goods. A family that builds simple, repeatable processes around cooking, preserving, and sewing becomes a micro-economy that can survive without outside funding.

A household that runs on low cash also preserves what money does come in. If you're paid in cash or if you barter for services, you'll want a simple ledger. Track inputs (hours, groceries, fuel) and outputs (repairs done, clothes mended, produce stored). The ledger isn't about blame; it's about clarity. It reveals where you rely on outside services, where your time is most valuable, and where you can save without sacrificing safety or health. The end result is not a perfect balance sheet but a practical map of your real dependencies.

In practice, this means choosing projects with short payback times. Replacing a worn-out tool with a durable, hand-powered version that can be repaired repeatedly is a smart move. It costs more upfront, but it survives multiple seasons of use, cutting future replacement costs and downtime. A well-run low-cash household organizes tools, spaces, and routines so that every action—cooking, cleaning, laundry, repairs—reduces the need for new inputs. The payoff isn't just money saved; it's resilience built into the daily rhythm of life.

Anecdotes from resilient households show the pattern clearly. One family repurposed an old bicycle into a hand-powered washing setup, dramatically reducing electricity use and turning the bike into a barter piece for neighbors who needed dry laundry. Another family kept a meticulous stock of staples and, during months of price spikes, traded extra jars of jam for a neighbor's eggs or a day of labor on a fence line. These are not grand gestures; they are small, repeatable moves that become social capital. When a household can consistently trade effort for goods or favors, its cash needs shrink, but its value to the community grows.

The final point in this section is simple: frugality without usefulness is a trap. Frugality that generates real goods, that creates durable options for meals, warmth, and cleanliness, builds a foundation for the rest of the system. A low-cash household isn't chasing nickels; it's building returns in time, skills, and relationships. That is how money becomes a tool, not a master.

GOODS PEOPLE WILL TRADE FOR

A barter economy isn't a dream; it's a practical channel for everyday value. The items that carry weight are those you can store, use, and guarantee in quality over time. Think durability, utility, and the calendar. The most reliable tradables fall into a few broad categories: food and preserves, handcrafts and repairs, seeds and seedlings, tools, and services that reduce someone else's time or risk. When you understand demand in your neighborhood, you can tailor production so it's both useful and saleable.

Food and preserves top the list because they fulfill basic needs and travel well. Jars of pickles, fruit preserves, dried herbs, and smoked meats are small, tangible assets that don't require refrigeration. Eggs, honey, beans, and flour are classic staples that many households can produce at home and trade for other goods. A family that cans 60 quarts of tomatoes each summer creates a reliable stock to barter through winter. It's not just about the product; it's about the guarantee that the product exists, is safe, and can be delivered on time. A handshake isn't enough if quality is inconsistent.

Barter goods extend beyond food. Durable clothing and blankets, mended shoes, and repaired tools are high-value items. A repaired shovel that works smoothly is worth more than a new one that wears out in a season. Simple, well-made wooden crates, tool handles, and hand-carved utensils travel well and hold value in a barter network. Seed saves and transplants are literally future value. A neighbor who saves and sells quality seed becomes priceless in spring. Seed packets are tiny but carry years of potential—if you trust the source and preserve the lines you know perform well in your climate.

The third category is home production. Soap, candles, laundry soap, and basic cleaning agents can be produced with common ingredients. Handmade clothing repairs are another strong card; a patched pair of trousers or a sturdy mending job keeps a family clothed without purchases. Craftsmanship creates value that outlasts any single season.

A well-constructed wooden box, a sturdy tote bag, or a set of kitchen utensils assembled from durable wood and metal can bridge shortages while remaining useful in ordinary times.

Finally, transport and energy-related items hold steady demand. Bicycles and carts, wheel repairs, and even the repair of a small stove or heater can save a household both time and money. People will trade labor for mobility—an hour of carpentry for a ride to a market, a day of repair work for a pass to a shared ride system, or a bag of fuel for a service you provide. The best tradables reduce someone else's risk or effort. They free time, reduce exposure to price shocks, and increase reliability.

You don't need to turn every product into coin. A reliable bartering partner is a person who remembers you when a trade is needed and who trusts the quality of your goods. Build that trust with consistent quality, honest pricing, and timely delivery. Your reputation becomes the currency you never have to print.

SKILLS THAT CARRY WEIGHT IN HARD TIMES

In disruption, skills matter more than flashy gear. The most valuable skills are the ones that keep a household moving with minimal outside inputs: repair, food processing, sewing, animal care, and practical transport. These aren't romantic or abstract; they're hands-on competencies you can teach a child and reuse for decades. Each skill reduces dependence on the grid, lowers costs, and increases the number of ways your family can contribute to the local economy.

Repair skills form the backbone of resilience. A broken latch, a loose door hinge, a cracked rain barrel—these are fixable with simple tools and a patient approach. The better you are at diagnosing a problem, the less you spend on replacements. That means maintaining a small toolkit: a hammer, a few screwdrivers, pliers, a hand saw, nails, a bit set, a square, and lubricant. With those, you can tackle most common faults around the house. A little rust removal, a sharpened blade, a readjustment to a belt—these small acts compound into a reliable home performance that outlasts the seasons.

Food processing is another essential skill set. Canning, fermenting, drying, and salting extend the shelf life of produce beyond what a refrigerator can offer. Canning requires a few jars, a good pot, and careful heat control; drying needs a rack or screen and a safe airflow. Fermentation needs clean vessels and time. The payoff isn't just safety; it's flavor, nutrition, and a buffer against sudden price spikes. A family that can safely store and later prepare meals from home-preserved ingredients will outlast those who rely on fresh purchases alone.

Sewing and textiles matter as soon as a store is out of stock or prices rise. A family that learns to mend worn clothing, sew patches on pants, or replace a zipper keeps wardrobes whole at a fraction of the cost of new items. It also teaches patience and physical control. The same patience applies to making durable household goods: curtains that last, towels that resist wear, blankets that stay warm after years of washing.

Animal care becomes an income and a food source when you have a small flock or herd. Chickens provide eggs and meat;

goats offer milk and compost; bees yield honey. Caring for animals isn't a hobby; it's a productive circle of waste reduction and value creation. Chickens turn kitchen scraps into protein, while goats convert brush into milk and fertilizer. You're not just keeping animals; you're building a farm-to-table loop that expands your household's resilience.

Transport is a practical, visible skill. The ability to move people and goods without relying on electricity or fuel is a huge advantage. Bicycles, handcarts, and wheeled platforms become lifelines in outages. Maintenance becomes a weekly habit: tire checks, chain lubrication, brake adjustments, and rust removal. Knowing how to repair a simple bike or wagon keeps you and your neighbors connected when buses disappear from schedules or fuel becomes scarce. A well-tuned bike is not a luxury; it's a small, stubborn lifeline.

Finally, the social dimension of skills matters. Teaching a child self-reliance, sharing repair tasks with a neighbor, or organizing a small repair circle makes skills contagious. A family that learns together keeps momentum. It's not enough to store knowledge; you must practice it, repeat it, then pass it on.

FAIR DEALING AND REPUTATION

The backbone of any working barter system is trust. Fair dealing isn't a moral plea; it's a practical system for reducing risk. When you trade, you set clear terms, meet deadlines, and stand by your promises. A reputation for reliability multiplies your options. People remember who delivered on time, who asked for too much, who botched a repair, and who showed up with a ready solution.

Clear terms begin with plain language. Write down what you will deliver, when, and under which conditions. If a job is contingent on a material you don't control, say so up front. There's no virtue in vague assurances. In disagreements, a ledger of transactions helps. Record what was agreed, what was exchanged, and when. A small notebook or a shared digital record can prevent sourness after a trade.

Price fairness matters as much as honesty. Don't chase market spikes with opportunistic jumps. The value of a trade should reflect effort, time, and risk. If a neighbor puts in a day of labor, you don't expect a traded egg to cover that cost. The best negotiators aim for mutual gain: both sides walk away with something valuable and no lingering resentment.

Reliability compounds. When you show up with the right item in good condition, you become the person others seek out when they cannot find what they need elsewhere. That reputation rewards you in kind—more favorable barters, shorter waits, and less pressure to barter away essential goods under duress. The money you don't spend becomes a social asset you accumulate through consistent behavior.

Disputes happen. When they do, resolve them with a calm, transparent process. Don't let a disagreement linger. A clear method for rectifying mistakes—replacing a damaged item, redoing a repair, or offering a alternative trade—keeps the network intact. In a tight market, your ability to absorb error and correct it without burning bridges is worth more than any single trade.

No household can endure long-term disruption in isolation. The strength of a barter network is the way it binds people to a shared capability: to produce, to deliver, to be counted on when the system around you falters. The better you perform in the small gestures of daily dealing, the sturdier your household's position becomes when a broader grid comes under stress.

This chapter has laid out a practical blueprint: find tradables that travel well, cultivate skills that turn waste into value, and build a reputation built on reliability. Do those things consistently, and your household will be known not for how little you spend, but for how much you reliably provide when others need it most.

FOURTEEN
MUTUAL AID, FAMILY ROLES, AND THE STRENGTH OF COMMUNITY

ASSIGNING HOUSEHOLD ROLES WITH CLARITY

No household keeps itself intact through a long outage without clear lines of responsibility. Roles aren't badges of authority; they're practical maps that keep decisions fast and actions steady when pressure rises. The key is simple: everyone knows what they own, and backups exist for every task.

Start with a simple map. List core areas: food and water, heat and shelter, sanitation, health and safety, tools and repairs, clothing and bedding, finance and shopping, and child or elder care. For each area, assign a lead who has ultimate responsibility, plus one or two backups who can step in on busy days or during illness. Put the names on a wall chart or laminated cards in the kitchen so they're visible at a glance.

Train in pairs. Cross-training isn't fancy; it's redundancy. If the cook is sick, the backup knows the recipe, knows where the stock sits, and can manage the stove and ovens. If the woodpile runner is unavailable, another capable person can move fuel, operate the hand tools, and keep the fire going. Practice a few simple handoffs so a missing person doesn't stall a task for hours.

Make the plan concrete. For each role, write down the exact actions that should happen in a typical week and in a grid-down week. Include trigger points—what signals a task needs to start (sunrise, when a pantry staple dips below a set level, or when a water container drops to a quarter full). Add emergency steps: what to do if a key person is away for a day, or if two tasks collide on the same person's plate.

Document and display. A one-page "roles and triggers" sheet, updated twice a year, keeps everyone aligned. Put it where the family naturally gathers: the kitchen, the mudroom, the barn. The goal isn't to micromanage; it's to remove guess-work when fatigue hits.

Put it to the test. Run a quarterly drill: pretend a power outage lasts a week, or a water issue closes a portion of the house. See who's ready to fill gaps, what tasks drift, and where the plan breaks. Those gaps aren't flaws; they're the system telling you where to tighten.

This clarity builds confidence. When the house has a rhythm and each person knows what comes next, the day's stress is less likely to derail essential work. The strongest households aren't busy; they're organized. And organization starts with naming roles, backing them up, and practicing until the routine is automatic.

In households that succeed, roles are not rigid cages. They're flexible, updating as people grow and as needs shift. A teenager who used to handle chores becomes a project manager for a repair season. A grandparent who rarely cooks leads the learning circle for preserving. Clear, shared roles keep family dignity intact when the lights go out.

End with a firm point: defined roles create reliable action, and reliable action is the backbone of any resilient home.

ORGANIZING SHARED LABOR

If a single household tries to do everything, the effort will collapse under its own weight. The Amish—and other resilient communities—lean on groups that share labor, not just goods. Shared labor multiplies capacity, reduces individual fatigue, and builds social ties that survive the hardest weeks.

Adopt a work-party mindset. Instead of scrambling to fix a broken fence, organize a two-hour session with neighbors. Bring a meal, build momentum, and finish the project together. In return, you'll have people who can lend a hand when your own backlog grows. The switch from lone effort to collective action changes the shape of the week.

Childcare swaps aren't a side perk; they're essential. A stable schedule of small, reciprocal childcare frees adults to harvest, preserve, or repair. It also reinforces trust among neighbors. When parents know the other households will look after their kids, they're more willing to take on risky or time-consuming tasks themselves.

Harvest help becomes a routine. In late summer, a schedule marks who's available for shared picking, threshing, or processing. A small group rotates through the work so no one bears too much of the burden. The payoff isn't only efficiency—it's community warmth and mutual accountability.

Repair days are the spine of durability. Set quarterly events focused on fixing, sharpening, and tuning essential gear: a wood stove, a windlass, a leaking trough, a cracked window. Bring tools, swap stories, and leave with a list of what to repair before the next storm. A tool-lending circle keeps equipment flowing without every home buying the same gear twice.

Document what you complete. A simple ledger records who contributed, what was fixed, and what's needed next. Clarity about labor avoids resentment and shows tangible progress when scarce resources press in.

Shared labor isn't charity. It's a system that keeps people moving when cash, fuel, and supplies tighten. It's also a clear signal to neighbors that you're present, capable, and trustworthy—exactly the traits that invite reciprocal support when it matters most.

End with a firm point: organized teamwork turns a single family's effort into a community's resilience.

BUILDING LOCAL SUPPORT NETWORKS NOW

A long blackout is social as much as it is physical. You can't sprint through weeks of disruption on stubbornness alone;

you need a web of local ties that can supply, borrow, and trade without the convenience of digital marketplaces.

Start with a map of talent. List what your block or cul-de-sac can offer: gardeners, carpenters, nurses, babysitters, seed savers, repairers, pet sitters. Then identify needs: who could use a weekly meal, who lacks a reliable water source, who needs a hand with clothes mending.

Bring people together with a practical plan. Host a low-key meeting where neighbors show skills on tables, benches, or a shared porch. Demonstrate canning, seed saving, simple first aid, or a DIY water filter. The aim is not to impress but to reveal the workable capabilities in the room.

Create a miniature local economy. Start a rotating "produce share" box, a seed swap shelf, or a repair exchange where people leave a note about what they need and what they offer. Establish a simple rule: no cash, no barter at all costs—trust, favors, and reliability earn priority.

Formalize a mutual aid circle. Decide how to respond if a neighbor loses electricity, if a family's well runs dry, or if someone ends up needing long-term care. Document contact information, primary and secondary responders, and a couple of fallback options. A printed plan, shared and revisited, keeps everyone on the same page.

Practice gives confidence. Run a dry-run of a weekend outage with a subset of the neighborhood. Track what works, what doesn't, and what resources are lacking. Use the results to shape a year-long rhythm: who to call, who to visit, what to bring for meals, and how to allocate scarce supplies.

A robust network isn't built overnight. It grows from repeated, small acts of reliability, from people showing up on time, and from the simple habit of sharing labor, tools, and know-how. The strongest communities aren't theoretical; they're practiced, visible, and trusted.

End with a firm point: real resilience rests on a web of neighbors who know each other well and act on it without hesitation.

RULES, TRUST, AND CONFLICT CONTROL

Disagreements will happen when resources thin and schedules tighten. The test is how quickly a group can reset, not how loudly a dispute is aired. A few guardrails keep conversations productive and prevent minor frictions from becoming lasting rifts.

Write a simple charter. Lock in decision-making methods, how to nominate a mediator, and how to approve big changes to the plan. Keep it short, clear, and revisable. Everyone signs, and the document sits in a common folder or on a visible wall.

Agree on consent-based decisions when possible. If consensus isn't quickly achievable, use a straightforward tie-breaker or a rotating decision-maker for that issue. The goal is progress, not perfect unanimity.

Maintain transparent accounting. A shared ledger tracks resources, labor, and money—no one feels excluded and no one suspects mischief. Regularly review the ledger aloud in a meeting so fault lines don't harden into secrets.

Set boundaries for time and space. Establish acceptable work hours, preferred dining times for shared meals, and a rule about borrowing gear that requires prompt return. Absence of a rule invites resentment; a simple rule protects relationships.

Create a conflict protocol. When tempers rise, pause, and revisit with a neutral observer present. If needed, rotate a designated mediator who isn't tied to a single household. The point is to cool heat before it becomes hostility.

Practice conflict management. Run a scenario workshop twice a year: two common disputes and the agreed resolution. Rehearsal reduces the chance of chaos when a real clash hits mid-crunch.

Trust isn't given; it's earned through reliability, transparency, and lived integrity. When people see that shared tasks get done, that promises are kept, and that the group can weather storms without bottlenecks, trust becomes the default posture—quiet, steady, and powerful.

End with a firm point: trust and clear rules aren't constraints; they're the framework that turns neighbors into a capable network when the grid goes quiet.

THE SOCIAL BACKBONE IN ACTION

The best preparedness plan isn't a gadget list; it's a social fabric you can lean on when systems go quiet. Mutual aid, family roles, and neighborhood labor create a home that keeps feeding, cleaning, heating, and healing—even when external support vanishes.

Consider a county that faced a weeks-long outage after a severe storm. Homes with strong mutual aid circles kept households fed through shared gardens and canned goods. A childcare swap let two families keep essential workers on the job. A repair party rewired a community shed to function without power, allowing a place to gather, cook, and stay warm. The outcome wasn't perfection; it was continuity. People kept showing up, and that consistency held neighborhoods together through the hardest miles.

Start small and practical. Gather a dozen neighbors, lay out a map of skills, and ask, "What can you reasonably contribute each month?" Then pick one shared project for the year: a neighbor-led canning schedule, a seed-saving circle, or a repair rotation for common tools. Make the project visible—photos, notes, and a running tally of progress.

Write down a simple plan for the next six months. Who will organize the first workshop? Who will host the seed swap? Who can lead a monthly pruning day? Capture commitments and post them prominently so no one feels left out.

If there's one truth in crisis planning, it's this: the grid failing doesn't just strain a single house. It tests a network. A well-ordered, trusted network reduces risk for everyone and creates the ordinary discipline that keeps life moving when supplies dry up.

AFTERWORD

THE STANDARD OF A RESILIENT HOME

By this point, preparedness should look less like a pile of supplies and more like a household that knows how to function. This book has pushed one idea from the first chapter onward: *systems keep people alive,* because systems keep working after the first rush of fear passes.

A resilient home has a clear standard. It can feed its people, secure water, cook simple meals, stay warm enough, manage waste, handle ordinary illness and injury, repair what breaks, and keep daily life organized when power, fuel, and outside services grow scarce.

That standard does not demand perfection. It asks for continuity, which is what most families actually need in a long disruption: steady meals, clean water, dry bedding, dependable routines, and people who know what to do when the easy option disappears.

Food sits at the center because hunger unravels judgment fast. A strong home stores staples it actually eats, grows what it can, preserves the harvest, and knows how to turn flour, beans, potatoes, eggs, broth, and canned goods into repeatable meals without leaning on refrigeration or convenience food.

Water deserves the same plain respect. Good enough means stored water in hand, a source beyond the tap, containers that can be moved by human effort, and a method to make doubtful water safer before it reaches the cup, the washbasin, or the cooking pot.

Heat and shelter matter because cold and damp grind down the body one hour at a time. A capable household can close off space, concentrate warmth, use bedding and clothing well, run safe backup heat, and rely on light sources that do not turn a hard season into a fire emergency.

Sanitation separates discomfort from sickness. When toilets, drains, laundry machines, and hot showers disappear, the household that can manage waste, wash clothes, clean hands, reuse water with care, and control odors holds onto dignity as well as health.

Medical readiness in this model stays grounded. It means calm attention to cuts, burns, fever, dehydration, infection, and daily health needs, along with the judgment to keep small problems from turning into bigger ones because the clinic is far away or overwhelmed.

Repair skills hold the whole structure together. A home with hand tools, spare parts, needles, thread, sharpened blades, patched clothing, maintained equipment, and people willing to mend before replacing stretches every dollar and every resource much farther than a home built on constant buying.

The same standard reaches outward into the yard, the cellar, the shed, and the neighborhood. Gardens, saved seed, preserved food, small livestock where practical, barterable goods, and trusted exchanges with neighbors all widen the margin between strain and failure.

The Amish example has mattered throughout this book for one reason: it shows what daily life looks like when resilience is built into ordinary living. Their households tend to keep more skills at home, waste less, repair more, depend less on distant systems, and lean on community before panic.

Most readers will never live exactly that way, and they do not need to. What they need is the same *working standard*: a home that can absorb shock without falling apart by supper, a family that can shift from convenience to effort without confusion, and a local circle that recognizes useful people when times tighten.

That is what good enough looks like. It is sturdy, unglamorous, and deeply practical, and it gives a family something rare in a fragile age: the ability to keep living on purpose when the grid goes quiet.

PRACTICE MAKES PREPAREDNESS REAL

Preparedness becomes trustworthy only after it has been used. Supplies sitting untouched on a shelf may comfort the mind, but practice shows whether the pantry supports real meals, whether the backup stove cooks evenly, whether stored water is enough, and whether the household can carry the work that crisis adds.

This is where many plans either harden into competence or collapse into fantasy. A hand pump that looked simple in the catalog, a lamp that smokes indoors, a pantry short on salt or fat, or a family that cannot find the can opener in the dark all teach faster than theory.

The strongest lesson in this book has been simple: start with the basics, then live with them long enough to learn their demands. Cook from your staple foods, wash with less water, use the manual tools, preserve what you grow, mend what tears, and see what breaks down first.

Ordinary practice changes how a household thinks. Dry beans stop being emergency food and become supper, flour becomes bread instead of an ingredient you vaguely hope to use one day, and cast iron, clotheslines, jars, buckets, root crops, and spare blankets take their place as daily tools rather than props in a preparedness story.

Testing also strips away false confidence. A family learns how much fuel disappears in a cold week, how heavy water feels by the fifth trip, how fast laundry piles up when it must be washed by hand, how much work a garden really asks in July, and how quickly tempers flare when routines are unclear.

That knowledge is a gift because it arrives while the stakes are still low. It gives you time to add another barrel, store more soap, sharpen the axe, patch the coat, improve the root cellar, simplify the menu, teach a child a task, or admit that one system needs a stronger backup.

Practice builds judgment as much as skill. You begin to see which chores must happen every day, which can wait, which tools earn their place, which purchases were wasteful, and which habits quietly save the household money, labor, and strain month after month.

It also forms family culture. Children who gather eggs, stack wood, wash jars, fold dried laundry, weed a bed, or help with cleanup learn that home runs on contribution, and adults stop carrying the whole burden in silence because work has names, owners, and rhythm.

The same truth applies beyond the fence line. Mutual aid works best among people who have already traded labor, shared knowledge, borrowed equipment responsibly, shown up on time, and built a record of keeping their word before the emergency puts everyone under pressure.

None of this requires dramatic drills or a survival costume. It requires repetition, honesty, and the willingness to let small inconveniences train you now so that larger disruptions do not teach the lesson at a harsher price.

The people who fare best in a long outage are rarely the ones with the most gear still in its packaging. They are the ones whose hands already know the motions, whose homes have already carried some of the load, and whose routines still hold when convenience drops away.

That kind of readiness feels quieter than the culture of panic buying and tactical talk. It is also far more dependable, because practiced habits keep working even when stress, fatigue, and uncertainty make everything else harder.

FIT THE PATTERN TO YOUR PLACE

The value of Amish-inspired resilience has never rested on copying another life down to the last detail. Its value lies in *principles that travel well*: fewer fragile dependencies, more household skill, a stronger local economy, durable goods, disciplined routines, and a habit of solving problems close to home.

That pattern can take many forms. A family on acreage may keep a large garden, a woodpile, a pressure canner, rabbits, chickens, and a workshop full of hand tools, while a suburban family may build resilience through pantry depth, backyard beds, a clothesline, stored water, careful repair habits, and strong ties with nearby households.

Even smaller homes can move in the same direction. An apartment or townhome may never support livestock or a root cellar, yet it can still hold staple foods, manual kitchen tools, backup light, warm bedding, basic medical supplies, repair materials, and a way to live with less electricity and less waste.

Climate shapes the details, but not the core. In colder country, heat, fuel, food storage, insulation, and winter clothing carry more weight; in hotter or drier country, water handling, shade, ventilation, spoilage control, and seasonal work rhythms demand closer attention.

Budget matters too, and this model respects that reality. Much of what the book has argued for—mending, sharpening, buying fewer but better goods, storing basic foods, saving seed, cooking from scratch, preserving harvests, using hand tools, and wasting less—often saves money while it builds strength.

That point deserves emphasis because many readers come to preparedness after seeing too much advice built around expensive gear. Amish-style resilience offers a harder and better bargain: less dependence on gadgets, more value from simple tools, and more return from skill, maintenance, thrift, and steady labor.

Age, health, and family structure change the shape of the work. One household may garden heavily and split its own wood, another may focus on pantry management, sewing, food preservation, local trade, and organized care work, yet both can become more stable, more useful, and less exposed to disruption.

The same flexibility applies to community. In some places, mutual aid grows among relatives and church families; in others, it grows through neighbors who swap produce, share tools, help with repairs, check on elders, or keep an eye on one another when services fail.

Adaptation keeps the model honest. A household that copies details it cannot support will tire quickly, while a household that chooses a workable mix of food storage, water security, backup cooking, warmth, sanitation, repair skill, and neighbor cooperation can sustain the effort year after year.

This is why the book has stayed practical rather than romantic. The Amish offer a useful pattern because they show how discipline, restraint, and local competence create resilience, but every reader has to translate that pattern into the realities of land, law, money, weather, health, and time.

Good preparedness always fits the ground it stands on. When the methods match the place, the work becomes repeatable, the systems become reliable, and the household stops chasing an image and starts building a life that can carry real weight.

CAPABILITY AS A WAY OF LIFE

After all the chapters on food, water, heat, sanitation, repair, clothing, seed, barter, livestock, and mutual aid, the deepest change is not found in any single shelf or tool. It is found in the kind of household these practices create: one with steadier nerves, clearer roles, stronger hands, and a better sense of what truly keeps life going.

That change matters because disruption punishes confusion. A family that already knows how to cook simple food, clean up without running water, mend a tear, manage a sick day, stretch supplies, share work, and ask for help from trusted people has already crossed a line that many households never prepare to cross.

Capability also changes the emotional life of a home. Anxiety has less room to spread when people can move from problem to task, when the pantry has depth, when water has a plan, when cold weather has a response, and when each person knows where he or she is useful.

This way of living carries a moral weight without requiring any reader to copy Amish religion or culture. It honors discipline, thrift, patience, honesty, craftsmanship, and neighborliness because those qualities make a house more livable in peace and more durable in hardship.

Useful people become anchors in unstable times. The household that can lend a hand pump, share seed, trade repair work, watch children, cook for an injured neighbor, loan a canner, or help bring in a harvest adds strength to the whole community and earns trust that no stockpile can buy.

That is why community has stood beside self-reliance throughout this book. Personal preparedness without dependable relationships stays brittle, while local cooperation without capable households has little to work with; real resilience grows where both are present and practiced.

The call at the end of this book is plain. Take what you have learned and put it into the shape of daily life: fill the pantry with foods you use, test the water plan, cook without the grid, repair what is worn, teach the children, clarify household roles, build ties with neighbors, and keep refining the weak spots you now know by name.

Do it steadily rather than dramatically. A family usually becomes resilient by adding one dependable habit after another, one repaired system after another, one season of learning after another, until preparedness stops feeling like a project and starts feeling like the normal way the house runs.

You will never finish in the sense that there is nothing left to improve. Tools wear out, seasons change, children grow, bodies age, prices rise, crops fail, and new pressures appear, which is exactly why a resilient household values habits over hype and character over display.

The promise at the heart of this book has been modest and strong from the start. A simpler, more disciplined, more locally rooted way of living gives ordinary people a better chance to endure long trouble with their health, dignity, and family order still intact.

When the current stops, habit takes over. Build the habits that keep a home fed, warm, clean, useful, and faithful to its people, and the silence of a failed grid will meet a house that still knows how to work.

Made in the USA
Monee, IL
04 November 2026

67383134R00095